# CONFESSIONS OF A SCIENTIST

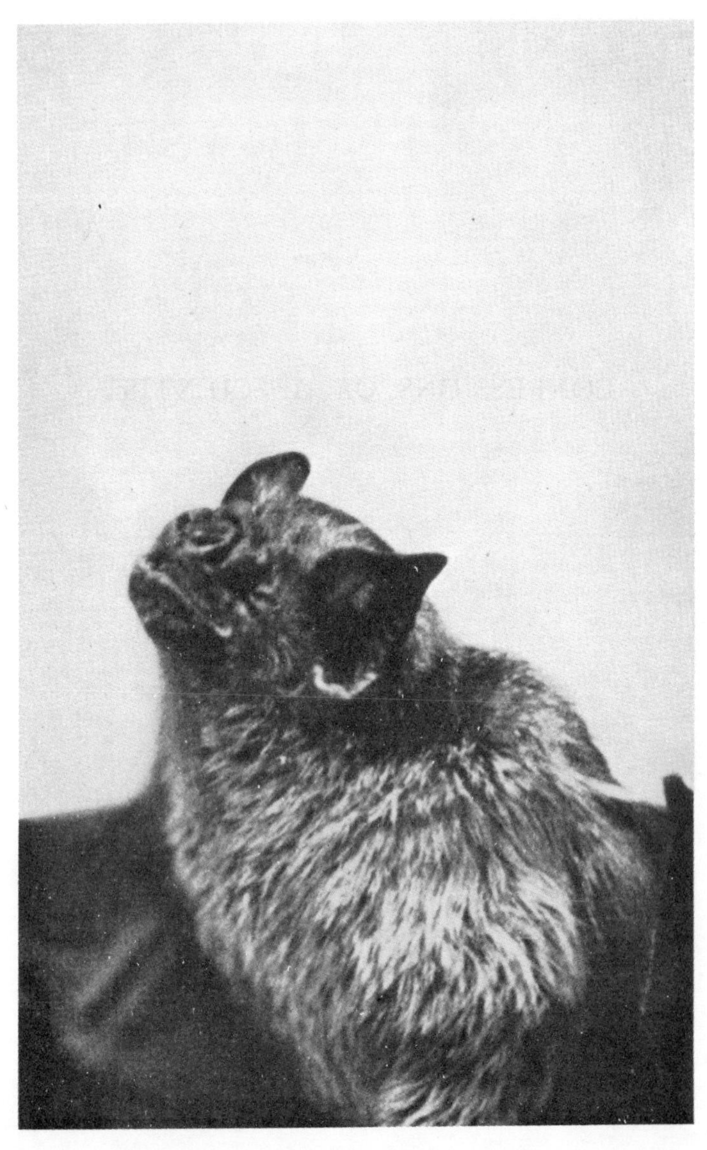

VAMPIRE BAT

The bulldog face with protruding lower jaw is characteristic. Sanguineous bats are known to occur only in tropical America.

# CONFESSIONS OF A SCIENTIST

*By*

RAYMOND LEE DITMARS

*Essay Index Reprint Series*

BOOKS FOR LIBRARIES PRESS
FREEPORT, NEW YORK

First Published 1934
Reprinted 1970

INTERNATIONAL STANDARD BOOK NUMBER:
0-8369-1800-2

LIBRARY OF CONGRESS CATALOG CARD NUMBER:
75-121463

PRINTED IN THE UNITED STATES OF AMERICA

TO GLORIA

## PREFACE

SEVERAL newspaper men had dropped into my office. It was late afternoon, in December. The days were short, and it was getting dark. We had been discussing a story that had several angles and my visitors had been scribbling memoranda. I am chaperon for a temperamental family, including elephants, lions, tigers, monkeys, kangaroos and wheezing, squeaking members of the small mammal section, but my office remains in the Reptile House. These quarters have gradually become invaded by sloths and bats.

I invited the visitors to remain to see the vampire bat drink her evening meal of blood brought from the city slaughter-house. Everyone appeared impressed. As the party broke up, with the necessity of my car making two trips to the lighted area outside the Park boundaries, there was unanimous insistence that I should summarize my experiences in a book. I told them that I had already done so. They insisted that my first attempts had only scratched the surface, that if I dug a bit deeper I should find plenty of other experiences well worth recording. I have thought it over and decided to confess some of the inner details of my work.

R. L. D.

# CONTENTS

# ILLUSTRATIONS

## ILLUSTRATIONS

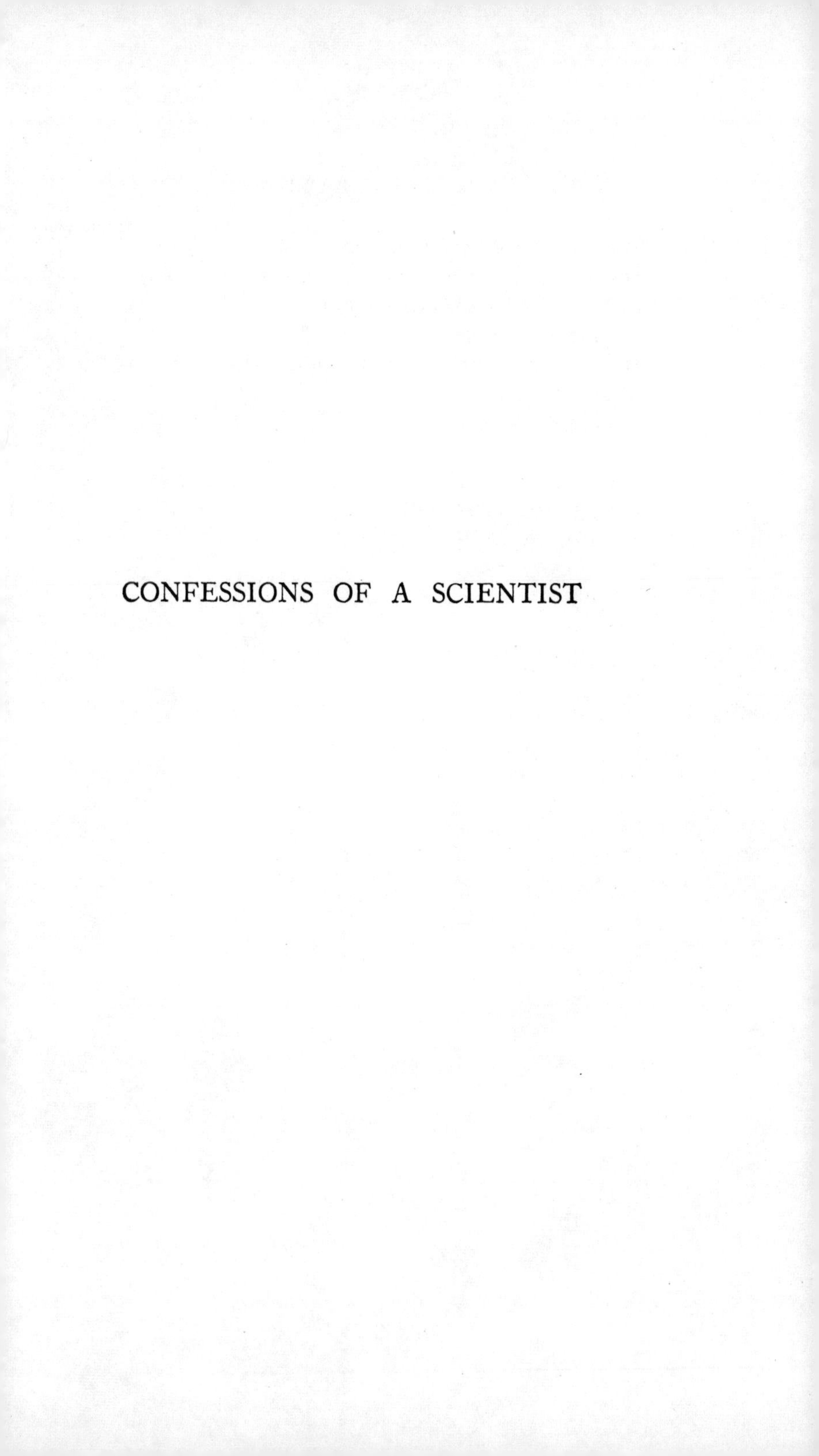

CONFESSIONS OF A SCIENTIST

CHAPTER I

# A FRUITLESS ADVENTURE

FRESHLY opened on my desk was what appeared to me an ominous letter. It read: "We are sending you a can of preserved specimens from a South American collector, and would appreciate your identification of this series." The letter was from a university.

The time was early August. I was clearing up matters preparatory to starting on my annual trip to the tropics. Nothing but superficial items had remained, and I had looked forward to slowing down, figuring on tropical clothes, a new kind of insect "dope" and other such interesting matters, in anticipation of breaking contact with my office. And here was an undefined quantity of pickled snakes, from what technical workers describe as "difficult" territory!

Well, that is how it goes in science. Things are pretty well checked off, when another puzzle is dropped in your lap. I worked on that can of preserved snakes for nearly two weeks. It was half the size of a barrel, and its odors crept through hallways and remained to greet me in the morning. The energetic collector, during weeks of time, had managed to corner the most puzzling array of diminutive reptilian forms. A number were bush snakes, which the scientist throws into two lots, the *Amblycephalinæ* and the *Dipsadormorphinæ.* That sounds formidable, and so it turns out to the softly swearing technician poring over ultra-scientific descriptions in search of the right title for the right specimen.

With that job completed, and after taking another jolt in the shape of a fat envelope of scientific proof, that had to be read before I left, I watched the casting off of hawsers from a vessel that broke contact with official matters.

Why was I going to the tropics? Didn't *that* mean scientific work? Yes, but I could do as I pleased, and choose problems that intrigued me. This time I was going to study the prowling grounds of the bushmaster, the largest poisonous serpent of the American tropics, the king of vipers, growing to twelve feet, clad in scales as rough as a wood rasp, the pinkish brown body with wide, velvety-black cross-bands precisely spaced—a striking pattern to be trailed across the jungle floor.

For years I had regarded the bushmaster as everywhere rare. In the thirty-three years' existence of the Zoölogical Park, we had received but three specimens, two of them from a scientist who lived in the tropics and spent his time prowling through jungles. Suddenly my mind-picture of the bushmaster had been changed by information coming from Panama. United States engineers had established a camp up the Chagres River. There was a five-year job ahead. The idea was to build a dam across the Chagres to hold back the floods of the rainy season, which sometimes disrupted ship traffic in the Panama Canal. Then, again, the big dam was to be useful during the dry season, as there had been times when there was not enough water to operate the locks properly. At the site of the operations a camp area had been hewn in virgin jungle. Here were houses on posts, screened against the night attacks of fever-transmitting mosquitoes. The building of the dam had been started, and the valley on each side of the river explored for a considerable distance to detect channels through which the backed-up water might get by. The contours of the valley were favorable, but a strange condition had been discovered. Deep under the rich soil from which magnificent forest-jungle towered, were great layers of limestone. In spots these protruded as reddish ledges, seamed with crevices and cave-like openings. They were not extensive enough to cause breaks in the forest, as the giant trees, many with a tendency to fling their massive boughs almost horizontally, shielded and shaded the outcrops.

Here was trouble for the engineers. Water worked past the underground, separated layers, and so for a considerable distance on each side of the dam deep borings had to be made, so that clay and cement could be pumped into the cave-riddled interior. Surveys for these borings radiated into the jungle, and in this work, during a bit over three months' time, eighteen bushmasters were encountered and killed. That was what altered my opinion about the bushmaster being everywhere a rare snake. It was also the reason for my trip to Panama.

Arriving in Panama I went to the Gorgas Memorial Laboratory, one of the world's most noted tropical research stations. My good friend, Dr. Herbert C. Clark, Director of the institution, had the preserved heads of the slain serpents. The Laboratory had offered a bounty on the heads of poisonous snakes killed in the area of the Madden Dam operations, in furthering a study of the prevalence of the different kinds encountered. The heads of those bushmasters ranged in size from close to the dimensions of a man's fist, down to a walnut, but the majority were large. They were blunt, with a ridge over the eye and jowls bulging upward toward the orbit. These features of the bushmaster's head give the effect of an ominous scowl. The upper jaws carry enormous poison fangs, teeth like heavy, hypodermic needles, over an inch long.

"The men up there can show you the very spots where they killed these snakes," said Dr. Clark. And with that, I was off to Madden Dam.

A few years earlier, a reconnoiter to the area occupied by the engineers' camp would have entailed some hardships. It would have taken several days to work up the river in a dugout alternately poled and paddled —a craft called a cayuco. It would have required camping materials and precautions against insect bites and fever, for which the valley was noted. I went up in an automobile, over a narrow cement road laid right through the jungle for a considerable part of the way. That road had to be built to get equipment and materials to Madden Dam. It is the gateway to a naturalist's paradise. It seemed almost unbelievable, when I reached the engineers' clearing in the jungle, to be given a room in the screened house of E. Sydney Randolph, the chief of operations, be shown a showerbath and told that the water from the taps was good to drink. One thinks of amoebic dysentery when drinking the water of such areas, but the engineers had constructed a reservoir equipped with chemical processes for purifying the water and filtering it.

A young engineer, Abe Halliday, was assigned to guide me over the surveys. In addition a native Panaman was provided for further penetration into the jungle. He turned out to be a good woodsman, and without him I might well have got lost. Arthur Greenhall, a scientific student of the University of Michigan, accompanied me as assistant, and was keenly alert and competent.

I had expected to find things pretty well torn up in spots, with the near-by jungle much disturbed. Such was not the case. Where men and machines had gone in on boring projects, the rapid, tropical growth was already reclaiming the trails, some of which were now difficult to follow. Halliday showed me the spots where several bushmasters had been killed. In each instance the terrain was similar. There were near-by limestone ledges, shattered or eroded with deep fissures. Some of the engineer's stories were hairraising. In all instances the poisonous serpent had been insolent and taken a stand, forming a lateral S-shaped loop in its neck and feeding more of its body into this loop, until the human observer, without realizing it, was subjected to great hazard by the increasing, striking length of the reptile. There had been flashes of the pinkish brown heads, and men had leaped backward, having a far less margin for avoiding the fangs than they had figured. In these encounters, Halliday explained, the tail of the serpent was in a blur of motion. This indicated a certain amount of fairness on the serpent's part. If the tail was among dried leaves it produced a loud buzz of warning. But if the forest floor was damp and mouldy, there was a barely perceptible sound. Bushmasters have a hollow horn capping the end of the tail, and this may rasp against dry objects; but it is solidly attached and does not rattle. It hints, however, of the ancestral stock from which the rattlesnakes were derived. Halliday told of one particularly terrifying experience. He had been sitting on a slope, waiting for his men to come up. There was a thick growth of

VALLEY OF THE CHAGRES

This river flows past many miles of virgin jungle. To the left is a giant ceiba tree. It was in this valley that the bushmaster was hunted.

WHERE A BUSHMASTER WAS ENCOUNTERED

The serpent edged out of the low tangle in the foreground and was shot by a surveying engineer.

THE SURINAM TOAD

*Above.* Found in northeasterly South America, this odd creature never leaves the water. Its eyes are minute, and it is as flat as a pancake. *Below.* From its odd markings the creature looks as if clad in a union suit. Note the enormously developed hind feet, for fast swimming.

ground-running vines behind him. There had been a shower and as usual, in the steamy period following, the dripping branches sounded everywhere on the leaves of surface vegetation. Through this, Halliday detected a buzzing in the leaves behind him. It sounded like a half-drowned beetle fighting its way to get free.

The sound was persistent and the engineer glanced around. He was horrified to see a pinkish head, with eyes of a muddy color, not a yard from him. He realized that he was within the zone of that bushmaster's stroke, and hurled himself down the slope like a rolling log.

"What happened to that one?" I queried.

Halliday heaved a sigh, as if the very memory of the occurrence came as a shock.

"He edged out where it was open. I emptied the clip of my automatic and hit him twice up near the neck. He was nine feet long. Dr. Clark has the head."

This was the kind of game we were after. It meant caution in climbing over the ledges from which bushy masses of vegetation protruded—fine lurking places. Halliday said the surveying parties seldom crossed the ledges, as there was no need of it. We decided that in exploring those ledges we had a better chance of seeing a bushmaster than men reconnoitering the floor of the jungle. Hence our hopes ran high.

How many miles of ledges and tangle we explored

that day is problematical, but on the second day we started in another direction. There was a spot, investigated on this trip, where one of the largest of the bushmasters had been killed. That spot warrants a few words of description. The snake killed was a few inches over nine feet long, and thus far appears to be the largest one encountered in the area. It was discovered in an opening on the floor of the forest produced by the dominating influence of an immense ceiba tree. At the base, this tree was close to thirty feet in diameter, its buttressed ridges flaring outward and curling together in places to produce cave-like interiors, which would have sheltered several men during a heavy rain. Above the buttressed portion the tree sloped inward like the shoulders of a gigantic bottle, and a smooth and round trunk a yard in diameter towered a hundred feet upwards before horizontal branches with thick foliage were outflung.

For a distance of a hundred feet in every direction the base of the forest was bare, as if the great tree ruled that no other growth should encroach upon its realm. In the background, however, were several massive quipos, which to me are the most curious trees in the tropics. They are alleged to have survived from prehistoric times. In form they are like massive, symmetrical columns, up to a yard in diameter and of impressive height. The particular point about them is a series of evenly spaced rings, about four feet apart on a big tree, protruding slightly and imparting the effect of smooth, cement cylinders, built to form a column.

"Well, here's where we shot him," said Halliday.

"And there's a limestone ledge!" was my reaction.

Forming a portion of the background was a particularly ominous looking mass of weathered stone. It was an ideal lurking place for snakes, and a spot to be examined with caution. It was broken into boulder-like fragments, and its uneven character rendered it possible for a coiled bushmaster to strike one in the thigh, or even above the waist. We used our ears and eyes in the hour's work of combing that ledge, but saw nothing of interest until we crawled into a cavern at the bottom. Here I saw some bats of surprising size. Their wingspread looked to be two feet across. They were so elusive during our efforts to capture them that we might have been passing our hands through moving shadows. Halliday remarked that some scientists had been after vampires in deep limestone caves near the river. We talked about outwitting bats with nets. That would be interesting to work out, and a vampire in the collection at the Park would be a high point of attraction. I had never heard of a vampire being scientifically exhibited. The thought of that sanguineous animal was recurrent during the rest of the day.

If the reader may think of any opportunities we omitted in hunting for a bushmaster in those jungles, I feel that the impression may be removed by the information that as we didn't discover a bushmaster during the daylight, we resumed the search at twilight and carried it into the night, with the aid of jacklights. This was eerie work. We went through trails and hollows to the clamor of frogs both on the ground and in the trees, saw many pairs of eyes reflecting in red or green the beams of our lights, but with one exception these turned out to be rabbits, fruit opossums or tropical whippoorwills. The exception was a small yaguarundi cat, which, to the detriment of romance, was sniffing around an empty tin can within the clearing of the camp. During all this exploring we had seen not a snake of any kind. Later, in investigating a quite different area where some cattle had been killed by poisonous serpents—which I suspected to be of the fer-de-lance species—I saw signs of the work of vampire bats among the herds. There were wounds in the hides of the animals from which "ropes" of dried blood descended. Here was further evidence of the inhabitants of the limestone caves, but the search for snakes although carried on for about two weeks was fruitless.

There was a most inappropriate climax to this expedition. Coming back to the Canal Zone we dressed up in tropical whites, visited scientific friends, packed our luggage and went aboard the trim passenger steamer that was to take us home. The jungle seemed far behind as passengers and visitors crowded the decks. Several of the stewards, whom I had met on former trips, greeted me as I sought my cabin. I had barely cast off my coat, preparatory to unstrapping a couple of satchels, when there was a knock at the door. I opened it to face a steward, who seemed disturbed.

"Can you come to the doctor's cabin, at once, sir?"

"What is the matter?"

"There's a snake among his books, sir. The doctor doesn't know what kind it is."

My first snake, after hunting for them for weeks in the tropics—and this in the cabin of a steamer! I hurried after the steward.

The ship's doctor sat calmly at his desk smoking a cigar. He greeted me and extended a hand toward his bookcase. Peering over the row of medical volumes was an innocuous, infant boa constrictor, about two feet long.

"Wasn't sure what it was," said the doctor. "Somebody brought it in here in that jar. It pushed the cover off."

I twined the boa around my hand and confessed to the doctor, that I hadn't seen a snake on my trip until this visit to his cabin.

A scientist has to cultivate a philosophic mind and accept failure. The engineers had not been looking for bushmasters, and they had killed eighteen. I had searched hard and had no success. The way in which chance plays a part was emphasized some time later when a visitor dropped into my office, a government man from the Canal Zone. He was on a visit to the States and he brought a bit of gossip from the area in which the vampire and the big carnivorous bats had been captured. The men at Madden Dam were on the watch for a bushmaster and were doing their best to locate one for me. There had been an attempt to capture one, and the tale of the exploit gave me a shiver.

Shortly after I had left with the bats, the young apothecary of the engineers' hospital situated in the jungle clearing was given permission to take the afternoon and evening off, and with another employee of the camp he started for Panama City for a festive evening. The two men left for the long ride in a light coupé, ran along the valley of the Chagres, through that area of magnificent jungle under the United States' care which is called the National Tropic Forest Reserve, thence into the Zone and to their destination. They spent a long and pleasant evening before starting back, with no chance of reaching the clearing at Madden Dam before two in the morning. There had been a steamy shower, and the monotony of the long ride was broken by watching frogs make prodigious leaps to avoid the car. It was the kind of night when the jungle creatures cross the road, and the two young men, both in their twenties, kept a watch for something of particular interest that might loom up in the glow of the headlights. Only a few nights before, one of the engineers had dodged a boa constrictor—not caring to try the experiment of running over it.

They were barely a mile from camp, when on rounding a turn they saw a big serpent stretched across the road. What impressed them was its pattern. It was pinkish brown, with wide cross-bands of black. They had come upon it too abruptly to stop or swerve and charged straight at it. What surprised them was the absence of any feeling that they had passed over anything. They stopped and backed up. Compressing the brake pedal to get a glow from the stop-light, the apothecary rose in the seat and glanced backward. The serpent had snapped its body into a coil, and the car had straddled it. The apothecary backed up farther, and once again the snake was in the glare of the headlights. There was no mistaking it now. It was a bushmaster, in majestic, fighting coil, and its tail was a blur of motion in a futile attempt to beat the characteristic tattoo on the road's hard surface!

Buoyed by the excitement of the find, the young apothecary jumped from the car, detoured around the serpent, and seized a long, stout branch. He made for the snake, maneuvered about it, got the branch over its head, pressed it down, then as the reptile started to writhe, pressed his knee over the branch for leverage, pulled on the bough with one hand and with the other reached forward and grasped the bushmaster by the neck. As he reached for it, bits of discussion he had heard at the hospital passed through his mind. "Right behind the head—so it can't turn," was one thought. "A fraction of an inch to spare, and it can use its fangs," was another. As he grasped the creature and hung on for life, he was sorry he had tried it. A whitish mouth flew open and long fangs gleamed in the light. The rough scales hurt his fingers and the creature appeared to turn within its skin. He let go of the branch and gripped the serpent's neck with both hands, but its rough-scaled body was now flung about in such fury that even a two-handed grip seemed insecure. The second man jumped into the struggle. Neither had ever handled a snake before, and here they were fighting a bushmaster in the confusing glare of auto lights. Tragedy hovered over that scene, as they struggled, panted and conferred about what they should do. It was the apothecary's idea to get the snake into the compartment at the back, which was built tight and close to carry supplies, and slam down the door.

In a scuffle, almost in a dance, they worked around to the back of the car. Here the serpent's struggles eased. It appeared to feel that it was conquered. A hand was spared to swing up the back of the car. At a signal given the creature was thrown in, and the top slammed down.

The camp was asleep when they drove in, so they locked the back, put the car in a shed, and decided that in the morning they would call a conference about caging that bushmaster and shipping him up to me.

The next morning, backed up by plenty of assistants ready with nooses on long poles, the top was raised with great caution, and the bushmaster was found to be dead. Its neck was broken. And this is a weakness of the big tropical vipers. The vertebrae of the neck are very delicate. Such serpents can deliver a lightning-like blow to imbed their fangs, but cannot endure restraint. The body was sent down to Dr. Clark at the Gorgas Memorial and stands as another tantalizing reminder of the area I hunted so persistently.

## CHAPTER II

# EPISODE OF THE TWENTY-FOUR TARANTULAS

Coming up through the Caribbean some details of my snakeless hunt spread about the ship. The doctor had probably dropped some hints about my only capture—in his cabin. The captain asked me if I would give a talk in the music room about my experiences. I accepted the invitation, but said the talk might be rather dull, as nothing much had happened. Events proved me to be not entirely correct.

Young Greenhall and I got together and debated about what I could use for illustrations. We had not come out of the jungle empty handed. We had some giant frogs, a "dragon" lizard which ran on its hind legs, some iguanas fully five feet long, and a particularly large and black tarantula. The latter I had noted crawling between my feet while seated on a log eating lunch. The iguanas, I must confess, had not been captured in the jungle. True enough we had tried to catch some of these large and elusive lizards, tried chasing them and snaring them, but without any luck. When we returned to the Canal Zone and walked out to the end of the breakwater, we saw a colony of iguanas running over the great blocks of cement. We had no idea of pursuing them, until we saw one run into a crevice between the blocks and leave its tail hanging out. When a moment later two others did the same thing, we discovered that the crevices only extended inward a bit over two feet and here was the finest iguana hunting to be imagined. We hiked back to Cristobal, obtained a packing case, attached a couple of rope handles and returned to the breakwater. With feelings akin to dreaming of the impossible, we shooed iguanas into crevices and selected the finest specimens. Really feeling sheepish about such an easy victory we took only a dozen.

As illustrations for the talk in the music room, we selected the largest iguana, the "dragon" lizard, the infant boa, one of the frogs, and the tarantula. Greenhall also brought out some of his preserved insects, including some enormous beetles with bodies as big as a goose egg. The two lizards and the boa were to be taken up to the music room in cloth bags. We obtained two large glass jars from the chief steward. In one of these we placed the giant frog, which looked as if it had been dusted with radiator gilding. The other jar held the tarantula. In order to make the big spider stand out prominently, I cut a circle of white cardboard as a floor for the jar. On this the tarantula loomed as a most sinister figure. We sallied to the music room and placed our exhibits in a row underneath the grand piano.

There was a fair crowd present, although the weather had turned bad. A tropical storm had gone by ahead of us, and the ship had developed a combination pitch and roll.

I outlined the quest for the bushmaster and described the capture of the "dragon" lizard. This creature was close to a yard long, with an excessively long tail. Standing above its head was a protuberance like the comb of a rooster. Greenhall had carefully removed it from the bag, holding it by the neck. Wishing to give the audience an idea of its speed in running, he released its neck and seized the tail. The lizard's long legs made a scampering rush along the piano, but the restraint on its tail caused it to leap into the air. There was a hurried movement of chairs backward, and my assistant, considering that the first illustration had been amply demonstrated, placed it back in the bag. In the meantime I explained that we had managed to capture this lizard by chasing it into a bush where it had become entangled. The audience registered the expression that we had done well in getting anywhere near such a whirlwind.

My next explanation was about capturing the iguanas and there was a hearty laugh about this, but it was suddenly hushed. I glanced at Greenhall. He had removed the big iguana from the bag. He was grasping it by the neck. Its mouth was wide open; and it was nearly as long as he was tall. It was a typical male of its species, with a row of high, red spines along its back, like decorations on an Indian's regalia. I sensed that our audience was praying that Greenhall would not give this creature a sprint along the piano as he had done with the first illustration. He took no such chances and my hearers settled down to listen to a discussion of capturing giant frogs at night, frogs which sing "boop, boop-adoop," with such clarity that their steamy jungles possibly furnished inspiration for the similar human renditions which have overwhelmed the world's concert halls in recent years. My description went on to the tarantula, and his jar was placed in the middle of the piano so that all might see this creature with its body and eight long limbs clad in lustrous black hair. My description was rather academic, to the effect that while the tarantula had large fangs and poison, it was not nearly so dangerous as alleged, and fatalities from the big spiders were almost unheard of. Moreover they never leaped at people to bite. They could run at considerable speed, and make short jumps—

There was a shuddering sigh from the audience. The ship had heaved and the tarantula's jar was sliding along the piano. Greenhall caught it, just before it toppled off.

I concluded my talk with remarks about the big

beetles and assured the audience that the tarantula would be shortly returned to a box covered with heavy wire netting.

As we were returning to our cabins with the specimens, Greenhall reminded me that there was a point I had not brought out about the beetles. An engineer at Madden Dam had claimed that in two instances the windshields of automobiles had been broken by such insects flying at night and attracted by the headlights. Our beetles were four inches long and a bit over two and a half inches broad. It was hard to estimate their original weight now they were dead and dried out, but their shelly covering was thick and hard. The engineer had declared that in one of the instances his car had been traveling about thirty miles an hour and he estimated that the insects could fly at close to that speed, so the impact was equivalent to an object striking at a speed of about sixty miles an hour.

I told Greenhall that a scientist had to be cautious in giving credence to such stories and repeating them without seeking personal proof. He came back at me by saying that if one couldn't relate such an incident about a tropical beetle without dashing around in a car trying to break a windshield, many interesting statements would have to be held in abeyance until proof had been obtained, frequently with great difficulty. As my young assistant was on the way to becoming a scientist, my attitude, of course, was to instill the spirit

GIANT BEETLES OF PANAMA

The male (with horns) and a female specimen are shown. It was declared that one of these hard-shelled, swift-flying insects cracked an automobile windshield.

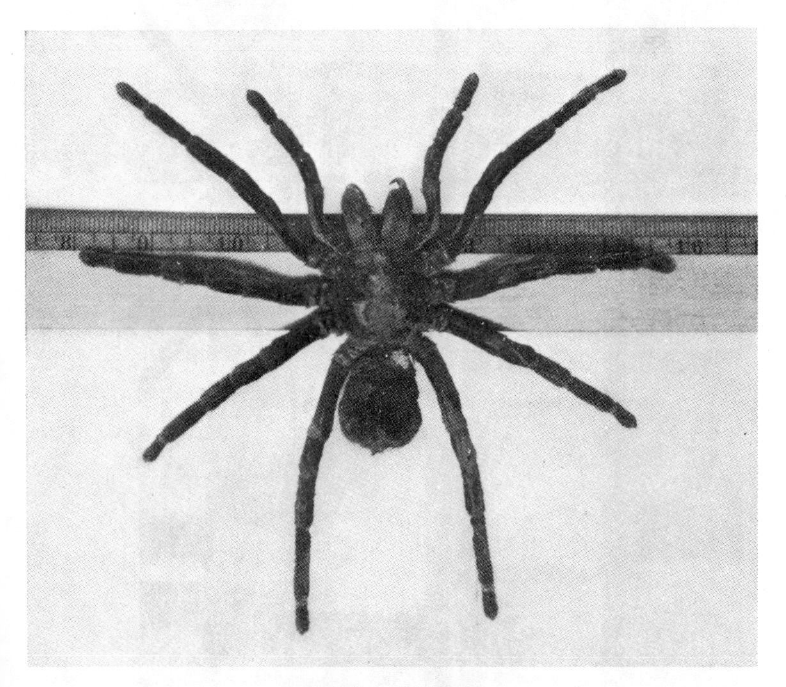

GIANT SPIDER

There is an eight-inch spread of limbs. The hook, showing on the right, close to the body, is one of the powerful poison fangs.

TARANTULAS FOR LABORATORY TESTS

Their poisons were studied during the development of the recent theory relating to the use of spider and serpent venoms in treating certain human ailments.

of caution at the beginning of a promising career.

At Havana I noted another of the trim white ships of the line coming in to an adjoining pier. On her bridge was the erect figure of her captain—and I recognized him. We had had a bit of a tilt the year before, although it ended in the most friendly spirit. It had been caused by tarantulas, none of which, however, was of the thoroughbred size of the one figuring in the music room episode.

When I had left for Panama the preceding year, Dr. Adolph Monae-Lesser of New York had asked me to bring back a batch of big spiders, as he wished to analyze their poison. Elsewhere in this book I have explained how his interest came from his observation of a leprous patient who had been bitten by a tarantula. The poison of the spider had appeared to have a certain specific effect in reducing the symptoms of leprosy. A man of Dr. Monae-Lesser's extreme caution in checking results, while obtaining data upon which to base a definite conviction, needed an ample supply of spiders. The poison reservoirs of a tarantula, situated at the bases of the joints carrying the claw-like fangs, contain a minute quantity. While these fangs are as large as those of a big rattlesnake, the orifice for the ejection of venom at the end of each fang and the connected poison gland are very much smaller. In relative proportions they might be compared to the tiniest wrist watch and an alarm clock. Smaller kinds of spiders appear to have relatively larger fang openings. Consequently these big spiders expend a mere trace of poison. However, it is very powerful. There is no doubt that if a tarantula expended as much venom as a fair-sized rattler, its bite would overwhelm a man within a few minutes, and kill him in less than an hour. Thus, drop for drop, if the infinitely small quantity of the spider can be mentioned in such terms, the tarantula's poison is more powerful than that of the snake.

Considering all this I obtained twenty-four tarantulas for the doctor. They were caged in a flat wooden box of twenty-four compartments, for these spiders are cannibalistic and have voracious appetites, and if I had placed them together their number might have dwindled one half every few days, until but one remained to represent an orgy of eating. I caught some grasshoppers in the salt marshes near the docks before sailing from Cristobal, and figured these would last me into Havana on the voyage north.

Within a day after leaving Panama trouble developed among the grasshoppers. They expired one after another. The curious thing about them was that they immediately shrivelled to mere traces. Possibly they were half starved in the first place, as the area where I caught them contained nothing but saw-grass, which caught in my trousers and looked dry and sunburned. I hastened to usher the survivors into the twenty-four dens of the tarantulas, but the way those creatures caused the food to fade away and waved hairy limbs for more gave me cause for worry. I went to the commander.

"Captain," I began, "you know that crate of tarantulas I am taking North. They have eaten all my grasshoppers!"

"What are you going to do?" asked the Captain as he stood watching the red flashes on the face of the radio-type fathometer. "Can't you feed them meat?"

"No. They want insects. I'm wondering if you would speak to one of the younger officers, to take me through the galleys, where there are roaches . . ."

The captain snapped off the switch of the fathometer, which, with each flash of the indicated depth over which we were passing, made a clattering sound. His attitude showed that he wanted no interruption while he gave me an answer. He swung around.

"Doctor," he said sternly, "I will assign the officer, but simply to convince you that there are no roaches on this ship!"

There was a strained atmosphere as I thanked him. This meant that the tarantulas fasted until I could go on a grasshopper hunt at Havana.

When I arrived at Cuba's famous port there were formalities ahead of me—delaying the capture of grasshoppers. Several scientists were to be visited, and this took time. I could not, of course, rush in upon these quiet gentlemen and tell them I must cut my visit short because it was necessary to go after grasshoppers. But the visits were concluded eventually, and there was a part of the afternoon remaining. I steered for the center of town to obtain an empty tin can, a large baking powder can or the like, for my insect quarry.

If one wishes to become acquainted with the psychology of Havana shopkeepers, it might be well to try to find an empty can. I was shown many cans, any of which would have answered, but all had something in them, though in a number the tide had fallen to less than an inch. In all instances the attitude was polite and cordial, but cans with falling tides could not be dispensed with. With the better part of an hour burned up I at last hit upon the simple idea of buying a full can, with the seal unbroken. This was effected without any trouble whatsoever.

Taking my can and walking down the Prado toward the ocean front, where there was meadow grass, I swung off to one side under the walls of an ancient fort. Crickets were singing in the crevices. This was hopeful. Crickets might more readily endure a sea voyage than grasshoppers.

At the base of the wall, I emptied the contents of the can. It was some kind of powdery, evaporated milk and a white cloud floated off. With my penknife I made a flap in the cover. A cricket was singing in front of me. With a long, stiff blade of grass I worked him out of his hole, when a voice overhead barked something in Spanish.

My knowledge of Spanish is meager. When I try to talk it, it comes in chunks and chokes me. Looking upward, I observed a soldier with a gun. He was not pointing it at me, but carrying it in a position suggestive of the "ready." It then occurred to me that Havana was in a state of tension coming from political troubles. It also occurred that my dumping of a pile of smoking white substance against the wall of a fortification might look curious to a sentry. There were voices, and an officer appeared. He was smartly attired and looked sensible. I held up the cricket and shouted "grillo."

That means cricket in Spanish. Then I dropped the insect in the hole in the can.

This did not appease the officer. He spoke Spanish in such rapid terms that I couldn't understand a word. It was evident that I was not to continue cricket hunting, so to emphasize that I would desist, I placed the can on a clump of grass and gave it a kick. The can went flying, and the cover clattered off to one side.

As I left the meadow, the officer followed me along the wall. I half expected to be arrested, but in dignified haste I reached the Prado and took a taxi to the ship. Playing safe, I related the incident to some Cuban officials in the dock office who spoke English. They told me not to be disturbed, although an elderly member looked stern and asked to see my landing card.

His eyes followed me as I walked to the gangway thinking of another plan.

I looked up my steward, a cheerful young cockney who was full of suggestions—not always helpful—and actively willing. I told him about my experience, and he made uncomplimentary remarks about the port, which stewards are inclined to do, except about the home ports. The tropic sky was darkening to purple, and it was all off for grasshoppers and crickets. Moreover, we sailed at eleven that evening, straight north for New York. I suggested that there might be roaches, probably of a generous size, in some place along the waterfront. If he could provide me with another can, and suggest a place—

This time his suggestions were intensely practical. He knew a place—it was along the waterfront toward the Ward Line piers. He had seen roaches crawl across the ceiling, run under the chairs—and he intimated their size. I deducted fifty per cent of the alleged proportions and decided the tarantulas would pounce upon such delectable quarry. He said the name of the place, in Spanish, meant the Idle Hour, or Pleasant Hour, or something of the kind, gave me directions to find it, and sallied forth for a large empty can, which, he declared he would garner from the store-room. He returned with the can and I started out.

By the time I had traversed a mile of the water front where cafés seemed to alternate with almost every other establishment, my steward's explanation as to what in Spanish meant Idle or Pleasant Hour was confused. All I remembered was his explanation that it was a "very nice" place, and seeing the white-ringed stacks of a Ward liner, and a crowded, convivial-looking establishment, I went in.

The proprietor inquired what kind of refreshment I might wish, and seeking to thus get acquainted I gave him an order. After being served I explained the nature of my quest. His reaction to my acquiring a collection of roaches was cold, almost antagonistic. But even while we were conversing I was staring at a roach of heroic proportions, which was negotiating the quadrangle formed by the top of an empty box for bottles.

He followed the direction of my stare, waved the fine insect from its promenade, and it streaked across the floor. I was becoming depressed with this struggle to feed tarantulas, when my taxicab driver of the late afternoon breezed in. He ordered beer, and I went over, joined him, and paid for the refreshments. I presented my problems, and told him that in the name of science I would stake him to a dollar, if he could induce the proprietor to be more cordial and permit me to obtain quarry for the can in one of the back storerooms.

He quickly changed the atmosphere and led me to understand that the nerves of persons in the city were jumpy; that there was a sort of surveillance directed from political headquarters, and anybody alleging that he wanted to snoop around for roaches was intruding on privacy with a strange kind of excuse. My taxi-driving friend grew voluble and nearly smothered me by the vigor with which he smoked cigarettes. "Can you blame the proprietor?" he asked. Who would expect a stranger to drop in and ask permission to search for roaches?

Thinking of the captain's attitude, I suggested that the proprietor didn't like the idea of my intimating that there were roaches in his establishment.

My friend waved the thought away. The proprietor had not been offended. How could he be expected to maintain a café in Havana and have a storeroom containing many boxes of bottles without a background of roaches? No, there was nothing to that. So we ordered more refreshments, and the proprietor was invited to join us. Time and considerable refreshments were consumed before the proprietor became cordial and we started for the storeroom.

My worry about the approach of sailing time and the possibility of obtaining enough insects in the limited period vanished in a moment. The proprietor went to a case containing some bottles in straw coverings. He gave it a thump with his foot. Insects of all sizes ranging up to the length of my little finger rushed out. We concentrated on the big ones. They leaped or were knocked into the deep can. Within ten minutes I had enough of them to feed the tarantulas on a longer voyage than that to the home port.

Aboard the vessel I covered the can with wire netting so that there was no possibility of escape, reached in with a pair of long insect forceps, and went through the process of extracting one roach apiece for the spiders. Each of the compartments had a hole too small for a spider to get through, but large enough for me to introduce the food and afterwards close it with a wooden plug. There was not a tarantula that didn't eat with relish.

The next morning I met the captain, and in a pleasant spirit he asked me if I had been successful in catching grasshoppers. I related the events of the afternoon—and evening. Once again he became formal, and asked to see the roach container. He looked through the netting and stiffened. As he turned to go, it was with a solemn warning that the escape of a single insect from that container would be a serious matter. I showed him the forceps and assured him that I had spent several years in technical work relating to insects and could handle them safely. I also told him that this particular species of roach would not thrive and multiply on a modern, steel ship. The last assurance didn't make a hit. He gave me to understand that the escape of one of those roaches would be highly detrimental to the reputation of the vessel.

When but a day from New York I told the captain

that everything was well. He appeared relieved and we had a pleasant chat. We had run into the westerly edge of a storm moving parallel to our course. By afternoon the ship was pitching enough to make walking difficult. A steward brought the captain's compliments and asked if I would come up to the bridge. There the captain showed me a marmoset monkey. He said he was taking it to a friend. His steward had been keeping it in the storeroom, but it had been eating very little. It had a belt about its body and a long cord was tied to a knob above the captain's desk.

"I think he's seasick," said the captain. "He wouldn't eat anything today and I had him brought up here. Can you suggest anything?"

Here was a chance for some friendly revenge. I had seen many marmosets half-starved in would-be kindly hands. And I had seen them go frantic when the proper food was offered. The point is that marmosets are not particularly fond of fruit and quickly tire of any one kind. Their favorite food consists of small lizards and insects.

"Do you want to see that monkey wake up and *eat?*" I asked the captain.

"I certainly do. What's the matter with him?"

"He's starving for the right kind of food!"

"He's had plenty of fresh bananas, and some grapes."

"Yes, and if he could talk and heard you now, he'd throw them at you."

"What does he want?"

"Insects!"

A gleam came into the captain's eyes—and he grinned.

"And you are going—" he began.

"To bring up the can of roaches."

I returned with the almost depopulated can and my forceps. An insect was extracted. As the marmoset leaped the length of the desk to seize the morsel it emitted a chatter of joy. It ate four of the big insects.

"Well?" was my query.

"Thanks," said the captain, "but I'm through with marmosets."

## CHAPTER III

# VAMPIRE

EARLY in 1933, when the annual urge to get down to the tropics began to grow, my uppermost thought was about capturing a vampire bat for the collection at the Park. I looked over records of the European zoos, records dating back over fifty years. There was no intimation that a living vampire had ever been exhibited in a zoölogical garden. The statistics of the great London Zoo were particularly complete. They extended back more than seventy-five years, and every species of mammal, bird and reptile exhibited during that time was listed, even to the scientific names. Asiatic and African fruit-bats had been shown in London, but no vampire, which is found only in tropical America.

Turning to the works of tropical explorers and to books on natural history based on observations gathered from various sources, I found that in reality little was known about the habits of vampire bats except the fact that they flew from their lairs at night to bite sleeping animals and human beings, that they appeared to feed altogether upon blood, and that the bites from their strangely developed teeth caused wounds to bleed profusely for hours. There were three species of these so-called bloodsucking bats. The quest for the vampire resulted. Young Arthur Greenhall again accompanied me.

Arriving at the Canal Zone and visiting Dr. Herbert C. Clark, Director of the Gorgas Memorial Laboratory, I found that a study of vampires was already under way. Dr. Clark and his associate, Dr. Lawrence H. Dunn, were demonstrating that while many bites of these animals among cattle and horses might be negligible in effect, certain individual vampires transmitted a blood organism fatal to livestock. These scientists, in their study of the blood parasites of tropical animals, had captured several bats and learned how to keep them alive. It was on their advice that we began the search of the Chilibrillo Caves, near the valley of the Chagres River.

Concerning the vampire we heard many strange tales. It was exceedingly sly. It ran like a rat over the cavern walls and darted into crevices. It attacked horses and sleeping human beings. It lived wholly on blood. It drank and rapidly digested amounts of blood out of all proportion to its size, making an incision with lance-like teeth, an incision so severe that bleeding continued for a long time.

In a shack near the caves we found confirmation of at least one of the stories. Here a boy had been bitten by vampires five times within a week, always on the under surface of his toes as he slept. He had bled profusely, and the earthen floor below his slatted bed was blood-stained each morning.

Scientists and shack dwellers united in giving the vampire a reputation disproportionate to its size. In reality the creature has a body barely four inches long and a wingspread of only twelve inches.

We established quarters in the middle of virgin jungle in the camp of the engineers constructing the Madden Dam, with the chief engineer, Sydney Randolph, as our host. From there on September third we set out on our vampire hunt. In the party were R. F. Olds of the Madden Dam staff, Sydney Randolph, Jr., and two young Panamans from Dr. Clark's laboratory.

The route led through cattle trails in the green, tropical tangle. The mud was ankle-deep and deeper, and the vegetation that brushed our bodies was infested with ticks and red bugs. It is fever terrain, and we were inclined to sidestep the occasional thatched native huts we encountered in the clearings, for perhaps seventy per cent of the inhabitants of the valley are infected with tropical malaria, which can deal a northerner a smiting blow. In one of these huts lived the boy who had been bitten by vampires.

The slope grew steep as we neared the caves, and a dense growth of rain forest made the going difficult. Our Panaman guides, pushing through barricades of vines, disclosed a hole in the ground, the entrance to the main cave. Disappointingly, it looked like little more than the entrance to a coal chute.

Adjusting the headbands of our jacklights, and with warnings from the guides not to break the wires attached to the battery boxes at our belts, we slid in and found ourselves in a horizontal tunnel in which we could walk upright in single file. The tunnel grew wider and higher. The floor was slippery with a pasty red mud. There was a roaring sound ahead, and we came to two big wells down which volumes of water were pouring.

Working past these cascades we were confronted with an underground stream, swift and knee-deep, that filled the tunnel from wall to wall. There was nothing to do but wade into it and push ahead. With the rush of water I felt wind. The flashlight showed that the movement of air was caused by the wings of two streams of big bats flying in opposite directions.

After splashing forward for a fair fraction of a mile, we heard another roar. It came from the source of this subterranean stream, which rushed from openings in the wall as from the gates of a dam. In a moment more we were again treading a slippery floor. The hallway was larger and now showed side galleries. The guides stopped to assemble the handles of the nets in which the bats were to be taken.

The atmosphere was unlike that of caves I had explored in the United States, where the air is cool, damp and lifeless. Here the air was hot, heavy and sweetish, with an effect dulling to the senses. Everywhere on the limestone walls enormous roaches crawled, fully three inches long, with waving antennae as long as my hand. Of a pale straw color, they glistened as if varnished. Here and there were huge spider-like creatures with a spread of limbs that would have covered a fair-sized plate.

The beam of the jacklight, directed upward, disclosed the flowing streams of bats, now twenty feet overhead. The crawling life on the walls was ominous. We respectfully kept our hands from too frequent contact with the stone, for in the cracks of these tropical caverns lives a bloodsucking insect that is suspected of carrying the organism of the formidable Chagas' fever, diagnosed and discovered by Dr. Carlos Chagas. The organism is a trypanosome, taking up its abode in numbers in the blood.

Now we entered a great chamber whose arched ceiling rose to an impressive height, in places perhaps to seventy-five feet, like the interior of a cathedral. The ceiling looked smooth, yet it was rough enough to provide hanging foothold for thousands of bats of a dozen different kinds. Each kind hung in a cluster of its own.

Some of the clusters were twenty feet in diameter; these were formed by bats with bodies the size of a

VAMPIRE BAT IN FLIGHT

There are but three known kinds of blood-drinking bats. All inhabit tropical America and have a wing-spread of less than a foot.

THE VAMPIRE WALKING

The author's studies indicate that the vampire walks as much as it flies. It is here approaching its nightly dish of defibrinated blood.

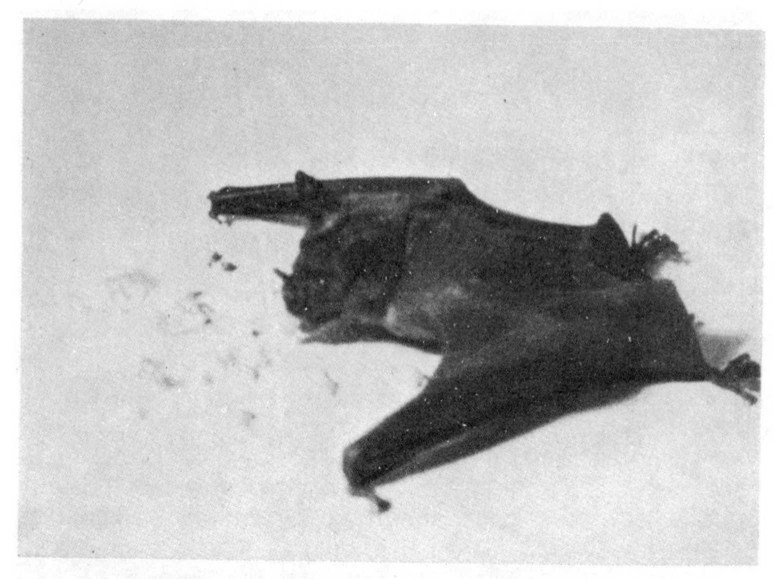

ORDINARY BAT POSITION ON GROUND

A big Spear-nosed Bat feeding. The average kind of bat "grovels" along the ground and must climb to a vertical surface to "take off" in flight.

VAMPIRE STALKING

In approaching its dish of blood the vampire is seen to assume the gait of a quadrupedal animal. From that position it can leap upward, and fly.

big rat. There were clusters a yard in diameter, made of bats the size of a mouse. The masses of big fellows looked like decorations of black crape. Our lights disturbed the bats, causing a great shuffling of wings, a peering of innumerable faces with flashing teeth, and a chattering like that of a legion of monkeys.

The side galleries also were full of bats, and we raided these in search of a big carnivorous species, which could not be captured in the domed chamber. We caught eighteen of that kind and fought them into a mesh cage. They had a spear-shaped appendage on the snout and their teeth were long and very white. In the light of our lamps they looked villainous. All the while we were watching for vampires, which can be distinguished by their habit of running along the vertical walls.

In working our way to one of the side galleries it was necessary to lie on our sides and edge along a mounded shoulder of limestone, skirting a black crevasse whose depth was indefinable. In the gallery we found bats of a new kind, but again no vampires. The dreaded crevasse had to be passed once more; then we retraced our way along the rushing stream until, to our relief, we saw a faint glow and realized that we were not far from the outside air and the daylight.

After a breathing spell, we sought and found the entrance to Cavern Number Two. The route sloped easily downward to a circular chamber fully a hundred feet in diameter, though not more than eight feet high. Here were hundreds of bats hanging in clusters and all of one kind, a medium-sized, spear-nosed bat of a fruit-eating species.

They were not timid, and we could approach close before they took flight. When a hand was waved the effect was of a pouring of winged bodies from the ceiling until the air was filled. We turned off our lights to discover whether the bats would strike us while flying. Although our faces were fanned with the movement of air from their wings, not one of us was touched by a bat in the darkness. Again in vain we searched the wall for vampires.

Cave Number Three had a sinister entrance, a well-like shaft. There was not room enough to get down with the nets and the party was weary. Greenhall and I lowered ourselves into the hole to a horizontal turn-off while the others rested outside, and, flashing our lights on the wall, we saw several bats run like rodents along the vertical surface, then dart into a cranny. Vampires at last! One of them turned for an instant and leered at us like a miniature bulldog.

With lights out we waited for half an hour. The vampires refused to reappear. We explored another gallery and came to a spot where a slender man might squeeze through, but we were too nearly worn out to continue. The only other passage sheered off at a ledge, and below us was a channel of water reaching from wall to wall, which looked as if it were over our heads in depth. There our day's reconnoiter ended. We agreed that a vampire bat was a prize indeed.

Back at the engineers' camp in the jungle, after we had washed off with a hose to rid ourselves of the red slime of the caves, we prepared to return to Cave Number Three at dawn on the following day. Engineer Olds helped us to make small nets with handles just long enough to grasp with one hand. Such a net could be swished instantly in any direction or slapped against a wall.

In the morning we were on our way while the jungle still dripped with the high, tropical humidity which builds up at night. At one of the thatched huts we picked up a Panaman lad who was lean and sinuous and seemed to match the tortuous alleys to be explored. With great caution we descended into the vampire grotto, keeping the lights out and feeling our way. Ready with our nets at the spot where the vampires had been seen, we flashed on the lights. There was not a sign of a bat.

Presumably, we reasoned, the vampires had retreated into the recesses of the tunnel with the deep water or into the narrow tunnel where a slender man might worm his way. To swim up the tunnel seemed impracticable, and the water looked far from healthful. Greenhall volunteered to work into the small horizontal shaft.

If the vampires were inside and there were no holes, perhaps he could net them. If they flew out past him, then I could have a chance at them. Light and slender, he writhed his way out of sight, though with strict instructions to keep within sound of my voice.

Ten minutes of occasional mumbled words, then he whispered, close to me: "There's several of them up ahead. The walls are solid. I'm going after them. Watch out with your net." A half hour passed before I heard him exclaim, "I have one!"

Meanwhile my light was illuminating the hole where he had gone in. A bat came winging in my direction and my net engulfed it; but it was not a vampire. Then I heard a heavy splash. "I'm in the water," Greenhall said, when I demanded to know what had happened, "and it's hard to get out."

"Come on," I bade the Panaman boy. "Get through to him." I gave him a lamp. The native lad went into the hole like a weasel. By the time I had got through after him, my belt had been torn from its loops and my shirt was around my shoulders. Luckily, the hole widened ahead.

Greenhall was wedged in a shallow crevasse and was waist-deep in water. There was no foothold by which to climb out. The Panaman youth straddled the crevasse and lifted my assistant back to the limestone shoulder from which he had slipped. Greenhall showed me the mesh box. It contained two vampires, one small and half drowned, the other an adult, a female, leering at us in fine vigor.

The adult vampire which survived illustrated how a scientist builds up the knowledge of a species, and the fascinating experience he enjoys in doing it.

Dr. Clark had demonstrated that a vampire cannot endure a fast of much over twenty-four hours. Also that these bats could be kept in vigorous health by feeding them defibrinated blood. The reason for defibrinating the blood is to keep a moderate quantity of it in a fluid condition. Vampires will not take blood which thickens or clots, as they are accustomed to drinking blood as it flows from a wound. To produce defibrinated blood it is necessary to obtain it fresh, which is easily done at a slaughter house. It is then subjected to a stirring or whipping with a bundle of straws, or, as some technicians prefer, with a bunched mass of beads. The straw or beads gather what is known in extracted blood as fibrin. It is this element which causes blood issuing from the carrying vessels to thicken or clot. Defibrinated blood standing in a tall bottle may show a precipitated darker red base to a moderate depth, but the great part of it looks like a brilliant red wine. The residue settling to the bottom may be stirred into the mass for use in feeding vampires, but no clotting tendency of the fluid bulk will result.

As we left for the Atlantic side of the Canal Zone, Dr. Clark gave me two quarts of defibrinated blood, fresh from the automatic refrigerator of his laboratory. From that moment, until we reached New York, the vampire became as intense a problem on our hands as if we were transporting the first-captured gorilla. We were naturally very keen to get it through. Its capture had excited lively interest in Panama and there had been local newspaper stories about it. I also understood that advice about the first vampire bat en route to a zoo had been cabled to the North. I was not worried about the eighteen big carnivorous bats. They were feeding ravenously, and meat could be obtained for them. But the single and much heralded vampire dominated our thoughts. With an assortment of crates containing reptiles and amphibians, and cases of preserved specimens for the museums, we boarded a train of the Panama Railroad for the Atlantic side of the Canal Zone. The blood was in a package beside me. The cage containing the vampire was swathed in black cloth so that she would not be shocked by the change from her eerie caves and the darkened jungles over which she had flown. As the train pulled out Dr. Clark cautioned me to get the blood on ice again as soon as I could.

On the Atlantic side it was necessary for me to stop two days at the Navy Submarine Base at Coco Solo. I was to deliver some lectures here, and the commanding officer had invited me to stay at his residence. We were driven direct to his attractive house and after an exchange of greetings, I might have been precipitate in requesting permission of his wife to place the blood in the residential refrigerator. The lady was hesitant until I assured her of its absolute healthfulness—for the vampire—and Dr. Clark's care in giving me nothing but the purest quality for the prize. The lady was gracious in extending permission and the vampire was housed in the garage. A quantity of blood was measured out in a flat, glass dish for it that night. The amount would have filled a fair-sized wine-glass. The bat hung head downward from the top of its cage when the dish was placed inside. It would not come down to drink the blood while I was there. Early the next morning we inspected the cage. The dish was empty.

That was the story during the ten days of voyaging to New York, with breaks in stopping at Colombian ports. We never saw the bat drink the blood, but in the quiet of the night she took her meal.

At the Park I decided to keep her in the Reptile House, which, owing to the automatically maintained temperature and the damp atmosphere, was like a greenhouse. Not only were temperature and humidity like those of her home country, but the sounds at night might have been rated to make her feel at home. The palms drip after the evening spraying, there are chirps and trills of tropical insects which have long inhabited the place, and every night there is quite a chorus from the "boop-a-doops", as the keepers call the big Panaman, bronze frogs.

She quickly settled down in commodious quarters. We defibrinated blood by our own laboratory processes, and judging from the way she consumed it she liked our methods. The dish was never placed in her quarters until it was dark, and as the days were still rather long my hours in returning home became erratic, and dinner was usually a stand-up meal gleaned from the ice-box. But I wanted to see her come down and drink the blood. She was becoming tame enough to show a lively interest when the dish was placed in the cage, crawled down a few steps, peered at it, then cautiously crept upward to her favorite nook in a corner, where she would hang head downward by one leg. Night after night she came farther down, and would wander along the vertical sides of the cage before again retreating. I was surprised at her deliberate motions, a slow stalk, head downward, and as deliberate a retreat. In spite of all this, however, I was unprepared for her astonishing actions when finally she did come down and go to the dish while I was watching. This happened two weeks after her arrival at the Park.

The blood had been placed in her cage and I took up my stand. There was a panel of roughened compoboard in the cage, to simulate the coarse limestone of the caves. She descended this deliberately, head downward, with ease and a certain grace. When she arrived at the bottom of the cage, her wings contracted until they resembled long, slender limbs. Actually, the wings of a bat are forelimbs, with greatly elongated "fingers" webbed with membranous skin. In all bats the "thumb" has a protruding claw used in climbing. In the vampire the "thumb" protrudes as a long fleshy member from the wing, on the end of which is the usual claw. This protruding "thumb", however, had been a puzzle to scientists. The hind legs of the vampire are almost free of the membranous wings.

As my vampire started across the floor of her quarters, the wings compactly folded so that they looked like slender forelimbs, she thrust her rear limbs straight downward. In this way her body was hunched and reared a full two inches from the floor. She looked like a huge spider, and her slow mode of progress increased the effect. Her long thumbs were directed forward and outward. They served as *feet!* To anyone not knowing what she was, there was no semblance of a bat, but a weird, stalking thing, of the softest gait. During her trip to the dish I realized that an unpublished habit of the vampire bat had been observed. Here, probably, was the method used in prowling over sleeping victims in search of a spot where the highly perfected teeth might start a flow of blood.

But other surprises were in store. Bending over the dish, she darted her tongue into the sanguineous meal. Her lips were never near the blood. The tongue was relatively long. It moved at the rate of about four darts a second. I had seen it at the instant of its first protrusion and it was a bluish pink. Once in action, it functioned so perfectly that a pulsating cylinder of blood spanned the gap between the surface of the fluid and the creature's lips. I had timed the start of the meal. In twenty minutes nothing but a red ring remained about the margin of the dish. The bat's body was so distended that it appeared spherical. She backed off from the dish, appeared to squat, and her wings spread like a flash. With a motion as light and silent as a piece of tissue wafted upward in a blast of heated air she left the floor, and in a movement so quick that I could not follow it had hooked a hind claw overhead and was hanging, head down, in her usual position of rest. Gorged, and inverted! It seemed a strange position after such a meal, but she preened herself like a cat, and stopped to peer at me intently in the light of the one shielded lamp to which she had become accustomed.

In less than half an hour, I had discovered some important things. Summing them up, I had learned that the vampire can assume a walking gait as agile as a four-legged animal, that the reason for the long thumb is its use as a foot, that it is not a blood-sucking creature as has long been alleged, that it can prodigiously gorge itself, and after consuming a fluid meal resumes an inverted position to digest it. Leaving the

VAMPIRE DRINKING BLOOD

These creatures consume and require a large amount of sanguineous fluid. The scene relates to the beginning of a meal consumed in about twenty minutes.

THE VAMPIRE COMPLETES THE MEAL

The action of the tongue is clearly seen. One point brought out in the tudies of the author shows the vampire is not a "blood-sucking" bat.

ALERT

A moment after the scene was flashed, the vampire leaped upward, and was in flight. The elongated "thumbs" on the ends of the folded wings, serve as feet.

THE LANCING TEETH

It is with the upper, center teeth, or incisors, forming a scoop-shaped lance, that the vampire instantly and painlessly produces wounds that bleed profusely.

vampire to her after-dinner siesta I started for my own belated meal, ruminating that what I had seen must now be recorded in motion pictures.

But the outlook was doubtful. If the vampire had been hesitant about performing up to this evening of exciting discoveries owing to a single, dim light burning, would she ever tolerate the powerful, photographic lights necessary for movies?

It took two more weeks of gradually developing the strength of the light before it could be done. Night after night I substituted stronger bulbs for the one first used for observation. For movie illumination I decided on three powerful actinic bulbs with a reflector. At the first trial of the photographic light, with a single bulb, she was much disturbed and would not come down. It was turned off and I waited for an hour in a dim light until she started her stalk. When she was halfway across the floor the light was flashed on. She hesitated, then flew to the ceiling of the cage. I waited close to another hour until she had returned and was at the dish. Then the light was turned on again. She finished half her meal, but was nervous, and left it. It called for patience, but her tolerance for the lights was carried through two, then three bulbs and the film records were made. They showed all the manifestations of the newly discovered habits. Owing to the necessarily moderate photographic light, and the impossibility of rehearsing a vampire to move about within the band of focus, there were some failures. The light was not strong enough for focal "correction" by reducing the size of the diaphragm within the lens. My eye was pressed against the focusing tube time and time again. I sought to split the focus in seeking for neutral spots where the bat might promenade, and hoped she would be reasonable. The first night's work came back from the laboratory marked "out of focus." I wanted needle-sharp definition in all the scenes. After two more nocturnal struggles I was beginning to get them, but of insufficient length. It was tiresome work, with constant shifting of the heavy apparatus. My assistant grew restive with the burning up of time and the skipping of dinner hours, but I told him we wouldn't quit until the scenes had all been developed at the laboratory and checked off as satisfactorily complete.

I learned a long time ago to take pictures while you can. You may think you're dealing with a healthy specimen and have all the time in the world, but a few disappointments teach you better.

I recall some giant walking sticks that a ship's officer brought me from the tropics. He fed them on lettuce, and they lived to reach me, but passed out soon after I had them in the studio and was getting ready to photograph them.

I had another experience with a little marsupial bear from Australia. We were the first zoölogical park in America to exhibit one, and I have described its arrival in another chapter. I had started taking pictures the day the bear arrived. Dr. Hornaday was director then, and he suggested that because the animal was pretty fagged out and on an irregular diet—graham crackers!—I had better put off the picture-taking for a few days. But I told him that I'd learned from experience to take pictures when picture-taking was possible, so he said to go ahead. It was fortunate I did so, for the "bear" died before an adequate supply of its usual food could be obtained. That experience was another that taught me to get pictures quickly.

But to return to the vampire. Barely were the pictures done when the most exciting event of all took place.

Just eighty-nine days after her capture, Head Keeper Toomey came into my office and made an announcement in the same calm tones with which he has communicated many and varied bits of information about the animals.

"The vampire bat has a baby," said Toomey.

"What!" was my exclamation.

"Sticking to her like a young opossum," remarked the head-keeper, as he continued on his rounds.

The portent of scientific discoveries never excites Toomey. I remember the first time that we had observed an African rhinoceros viper shedding its skin and it emerged in hues that would have dimmed the pattern of an oriental rug. It disclosed symmetrical squares and rhombs of olive, carmine and blue. We gasped as the amazing creature rolled back the tissue of epidermis which had hidden the brilliant colors.

"Makes a butterfly look sick," was Toomey's remark as he strolled off to consult the chart of the recording thermometer.

Toomey's advice about the baby vampire sent me on the run to the cage. An almost hairless object, like a little mouse, with folded, pinkish wings, was clinging to the mother's breast.

In more than thirty years with the big family at the Park, I can remember no specimen that afforded me more interest than that tiny bat in its actions and development. Motion pictures were impossible, for at any action in front of the cage the mother would hide the baby under her wing, or drive it into a sheltering crevice in the upper framework, where it could not be seen.

Imagine our shock twenty-four days after the birth of the infant, which had in the meantime doubled in size, when we found the mother suspended in a natural position at the top of the cage, but dead. A critical examination revealed nothing the matter with her. She appeared to be perfectly nourished, her muscles firm and strong and her silvery coat smooth and lustrous.

The restless infant, squeaking hungrily, and prowling up and down the sides of the cage was taken in hand by Keepers Rimmer and Taggart. These men have for years been accustomed to rearing strange and delicate types. Rimmer, for instance, discovered that juices produced by boiling down different kinds of fruits are highly beneficial in rearing young monkeys, while Taggart is an expert in the collecting of varieties of soft-bodied insects for extremely small, scientifically valuable specimens which will take nothing else. With the sympathetic interest of these men focused on the infant's welfare, we had a conference and decided that the young vampire was developed enough to digest the defibrinated blood, if it was thinned by filtering.

The little bat substantiated our decision. It drank the fluid with the same darting of the tongue we had observed in the parent. It was given small quantities, a fair portion of a teaspoon, every three hours during the day, or four such meals a day. On the second day after its mother's death it quickly came to Rimmer's hand to take its meals. As on the third day it continued to drink and digest the blood, we had high hopes of rearing it.

On the morning of the fourth day, however, we found it hanging dead at the top of the cage.

Two days after it expired, I gave a scientific lecture at which the vampire films were first and formally presented. Glancing toward the picture, at the enormously magnified mother stalking across twenty feet of screen like a spider of colossal proportions, the thrill of outlining an original, scientific discovery seemed to fall rather flat. True enough, an important study had been accomplished, but despite the ugly, wrinkled face and sanguineous habits of that vampire, I had developed a real affection for her. I have been told that this is not a proper temperament for a scientist, who should jot down cold facts and pass on to another job. I will confess, however, that I have not followed that practice, and there may have been times—even with the bat—when thoughts of giving an animal a rough deal have caused gaps in my notes.

Then again, there is Head Keeper Toomey's point of view. As the term vampire has a thrill for the newspapers, the capture of the bat, the first exhibition of the species in a zoo, and the birth of the young vampire were closely followed. The captive became famous from coast to coast. There were many stories about it. As the mother and her baby were finally placed in repose, side by side, in a museum jar, Toomey made a remark which summed up the stories and the movies, and the way we had lurked around the cage making notes for a scientific article.

"Those bats certainly worked for their keep," said Toomey.

CHAPTER IV

# INTRODUCING THE SCIENTISTS

THE preparation of a scientific article about the vampire now confronted me. I approached this with the greatest care, examining the bibliography to find if there had been somewhere, or at sometime, descriptions relating to the strange walking of the bat, the use of the thumb as a foot, the method of drinking, not sucking blood, the birth of the young, and its development. The preparation of that article extended through weeks of intermittent writing. In addition, young Greenhall started on a thorough search for references to vampires in the great library of the University of Michigan.

The reader may wonder what all this fuss was about, why I didn't go ahead and write my article. The answer is that no scientist desirous of retaining his standing in the various clans goes precipitately into such things. He has too much respect (I might even say fear) for criticism of his claims for priority in discovery. Such criticism may appear in printed form, or travel the "underground" way, by word of mouth. When I mention the clans I am referring to the specialists who study the warm-blooded, four-legged animals, the birds and the reptiles. They are formally designated mammalogists, ornithologists and herpetologists. There are many other groupings of scientists, of course, but the work of the three mentioned relates to zoology.

Occasionally disgruntled technicians speak of scientists as inclined to be "catty," and declare that there is much professional jealousy among them. This feeling is usually the result of unfavorable reviews of articles, or marked difference of opinion. The "catty" designation is unfair—except in occasional instances. As for professional jealousy, there is reason for it. Scientists are justly jealous of studies they have developed, and resent rash assertions in word or writing which may divert credit from the rightful place. It is hard to "get away" with anything among modern scientists, and this means that the weaving of the scientific fabric contains few flaws. In such an atmosphere the scientific review of allegedly new habits to be added to the history of the vampire had to be checked and rechecked as it developed.

I remember an article of the kind that had to be readjusted after it was written to admit the findings of another scientist which altered the picture. I had obtained a pair of rare South American monkeys, mere miniatures no larger than squirrels, of a supposedly delicate type. Records from other gardens indicated that they had an average life in captivity of about six months. These marmosets were beautiful creatures, with long, silky hair platinum blonde in color. I had watched wild examples in South America, and had ideas about maintaining captive specimens. They must have plenty of sun and a variety of food, including insects.

Their commodious cage was placed in a double southerly window. Early in the day the marmosets were provided with a tray of mixed fruits, and in the afternoon they were given a meal of grasshoppers. The latter meal was the more ravenously devoured. When winter came on, and insects were scarce, I substituted small lizards, a series of which was maintained to feed certain kinds of tree snakes, which will take nothing else.

The pair of golden marmosets thrived and broke records of life in captivity. Then an interesting event occurred. Twins were born.

The infants were large as compared to the mother. They were the size of a third-grown rat, and clad in golden fluff.

Development was rapid. They nursed vigorously, and their hair grew longer. When a week old they wound body and tail around the mother's body, encircling her like a broad belt. There were regular intervals of nursing, an hour or more apart. Up to this time, the father, though occasionally pawing through their silky hair, paid little attention to them.

Then came a sudden change in parental attitudes. I noticed that the father was yanking at the babies. He was rough about it, and I was inclined to remove him from the cage. Later, I returned to watch, and was perplexed to find that the *father* was now carrying the infants enfolded in his arms, in the position that had been assumed by the mother. Here was a predicament. How were they to be transferred for nursing? Also, was the mother losing interest in them? I decided to let a few hours go by, and see what happened when infant appetites were thoroughly aroused.

Just how long the father had been carrying the infants I was in doubt. The transference had occurred while I was not watching. After an hour of peering on and off around a corner, I made a surprising observation. The mother was sitting on the floor of the cage picking through some hay. Suddenly the male leaped down beside her. With vigorous efforts he divested himself of the clinging infants. There was much squeaking and complaining on their part, but his actions were unmistakable. They crawled from the father to the mother and started to nurse. The father sat quietly by. The proper period of obtaining nourishment expired and the parents reacted simultaneously. The mother stretched. The father yanked at the infants and despite their squeals kept at it until they were again in his arms. The whole action was so definite, in relief of the mother, that I felt it would be repeated; so I stationed a keeper to watch the monkeys. He reported that less than two hours later the performance was precisely duplicated. Things were carried on in this way for weeks, the father taking care of the babies except at periods of nursing.

Within a few days after the new system started, the young monkeys would immediately leave the father for the mother, and as quickly return after they had nursed. The news got about the Park, and the cage of the golden marmosets became a center of attraction. A newspaper took up the story and stated that here was a shining example for the husband who grumbled about going out with the baby carriage on Sunday mornings.

Paternal solicitude developed to a point that appeared pathetic. Within three months the young marmosets were a third grown, but still clung to their father. They were now of such a size that they sat upon his back and bowed him down, but there was no hint of impatience on his part. Even after they were only nursing intermittently and nibbling at tender bits of food, they half smothered him.

Once the promising twins were steadily eating varied food of their own accord, a new phase of family discipline developed. Both mother and father had been ravenous when big grasshoppers were placed in the cage. All consideration for each other was cast aside. They would fight for the insects. Now there was a change. When we offered grasshoppers to the young marmosets and the mother piggishly leaped for the morsels, the father drove her off, grasped the insects, bit off the heads and tendered them to his progeny before he himself took any.

From this picture it would seem safe to write an article declaring that the male of the species of golden marmoset plays an important part in caring for the young, as has been observed to be the case with some of the large, wingless birds. The male emu, for instance, incubates the eggs and rears the young.

Searches of the records, however, disclosed the studies of a scientist indicating that his observations among other kinds of marmosets pointed to paternal care being confined to the *younger* males, this coming from their tendency toward "social" habits. This scientist cold-bloodedly outlined his conviction—based on a series of observations—that "social" habits among marmosets were actuated by the selfish thought of bossy young males that the progeny would pick through their hair for imaginary fleas, a prevailing diversion among monkeys. Thus these young males had a failing for a condition akin to getting their backs scratched, and took pains to initiate the young into satisfactorily doing this very thing. In thus taking the punch out of my interesting observations, the scientist declared that not only young fathers would carry immature marmosets about with them, but young bachelor males "borrowed" the young from mothers. It seems that the only redeeming point left was the return of the young to be nursed. Old marmosets had no use for such nonsense, and consequently a spouse with an elderly husband had no hope of going out and leaving the children with the father.

What is the personnel of the zoologists, who work along these lines? What do they look like? In a convention—and such gatherings are frequent—there is no difference in their looks from a gathering of business men. Conversation, of course, is materially different. Papers that are read are not always of general interest to the audience, but there is seldom a paper which does not in some way broaden some phase of research somebody else is following. I remember a meeting where a member chalked on a blackboard his suggested "key" to a group of turtles which had been the cause of dissension for years. I happened to be working on those same turtles.

A "key" is a grouping of species into several sections. There may be major groups A and B. In this are subdivisions under *a*, *aa*, *b*, *bb*, and the like. Under A (if the key relates to turtles) may be the designation, "Shell but slightly longer than broad." Under B "Shell much longer than broad." In the subdivisions are short descriptions of colors and names of the species. To define species from a key is fascinating, like working out a puzzle.

In the turtle key on the blackboard there were two groups of species separated according to the number of claws on the hind foot. One group had four claws and the other three. It looked to me as if this difficult grouping of turtles would be set up as readily defined species. I had determined, at any rate, to similarly recognize them, after acquiring elaborate material except from one source. I conferred with the scientist, and our ideas were in harmony. He asked me if I had ample specimens, and if my "localities" had been exactly tabulated. I assured him of this. Little did either of us think that these very points were to put dynamite under our theories.

I had written to my Florida collector to send me a very full set of specimens. The turtles arrived, and with them charts the collector sent me showing the spots where the turtles were captured. If somebody comes across them in later years, they may send the finder on a search for hidden treasure.

In shell characters the turtles worked out all right. In the notch, or hook in the upper mandible, they also worked out. The examination extended to the claws on the hind feet and I came on one with four claws where there should have been three!

"Remarkable abnormality on the left side," I murmured.

Hardly able to tear my eyes away from this arbitrary foot, I looked at the other. It also had four claws!

Much disturbed I reached for turtles in all directions. I found others with four claws, and some were among those I had shoved to one side as belonging to another species. I found three claws on a few which should have been four-toed individuals!

Packing the whole outfit, I sent the turtles to my colleague who had aspired to present a "key" of the difficult genus. I soon received a long and very sad letter. The inconsistency in claws had blown his key sky high. Together we turned to shell characters and decided to give the species a good rating on that basis. They continue to stand, but I fear they are like the pins in a bowling alley, tempting some future scientist to take a shot at them.

In these references to scientists, the reader may query the whereabouts of the "typical" scientist, the alleged absent-minded character absorbed in his problems, who forgets to eat and does other eccentric things. This type has figured in fictional books and has been much overdone. There have been scientists like this, but they belong to the older generation and there are not many of them left. I met some of them when I was much younger, and remember one professor of the science department of a university, who spent all his spare time during the day, and much of the night, dissecting the head of a shark. It was always the same head. This went on for well over a year. In his room was a jar the size of a tub, and in this the shark's head reposed between times. On a table was a big metal tray, and while he was working, the head was placed in this. The dissecting tools were the smallest of scalpels, forceps and probes. I enjoyed the professor's confidence to the extent that I could sit beside him and watch him work. His progress was by fractions of inches, and he kept a lens like a jeweler's glass plastered to his eye. He never told me just what he was doing, and when others came into his study he would draw a towel over the shark's head. The aroma of preserving alcohol pervaded his room, and in this atmosphere he lived, pale and intense upon his studies. Frequently a sandwich or some buns and a cup of coffee growing cold stood beside his odoriferous problem. Nobody ever learned what he sought. He died during an epidemic of grippe. The shark's head, with its many little craters of cutting and probing, showed nothing understandable, and his voluminous notes were of a cryptic nature beyond explanation.

Such labors may seem impractical to the layman, considering the great consumption of a man's time; but many of them, when combined with the results obtained by other workers, produce discoveries of great value to mankind. They clear up points relating to the workings of intricate parts of the body, the functions of glands, the activities of certain nerve centers, the *reason* for puzzling manifestations, and the like.

Occasionally one does meet the extremely eccentric

type of scientist nowadays, and he may not, strictly speaking, be a member of the old school. He may appear so absent-minded that he walks into posts and does other erratic things, but analyze his character and you will find him far from needing a chaperone. I have a very dear friend like this, an entomologist. He is about fifty years old and of very slight build, but he can go on hikes through thick places which would bring the average man almost to the point of exhaustion.

I was working on some studies in an institution where the entomologist was one of the party. There were four of us, including a middle-aged woman of the library staff who had brought in the loose-leaf files for reference.

The professor had given us an example of his absent-mindedness. He had come to my table and borrowed a screw-driver from a box containing some simple tools. He wanted to pry off the top of a beetle case. Later I wanted to do the same thing with a case on my table. Opening the box I looked for the screw-driver, but found a pair of scissors.

"Who has the screw-driver?" I asked.

"Returned," mumbled the professor as he held a pin between thumb and forefinger and squinted at an impaled beetle.

"No screw-driver here. There's a pair of long scissors."

"Yes, yes. Just so," admitted the professor sheepishly. "I knew I returned something." He handed me the screw-driver.

We worked past the regular hours, and the lights of the big, departmental room were turned on. The lady member of our party suggested that she would like some tea, and produced an electric contrivance like a percolator. When the prongs attached to the cord of the percolator were pressed into the wall-plug there was a flash. The lights went out. There was something wrong with the tea-brewing affair and a fuse had been blown.

Our first thought was to telephone the building superintendent to send an electrician upstairs, and somebody groped for the telephone. There was no answer. To go to the basement and search for somebody meant walking through a fair fraction of a mile of exhibition halls and down many stairs, as the elevators had stopped running. The lights were burning in a long hall outside, and we went in search of the switchboard. It was a panel a yard high, filled with a maze of cartridge fuses, little red cylinders, all of which looked alike. With some fuses when a line has burned out, a circle on the side shows black. With these fuses, there was no such tell-tale marking. The thing was a puzzle. Extra fuses lay on a ledge at the bottom, but how was one to find the circuit of our darkened room?

A change flashed over the professor.

"Robert," he snapped to our young assistant, "go back to the room and call to me when the lights come on."

He turned to the librarian. "A hair-pin, please," he said.

He was handed one of a slender variety, and shook his head in disparagement at the degeneration of the article from the stout accessories of the past. But he bent it into a wide U.

With this he spanned the fuses, the tips of the hair-pin touching the prongs holding each end of them. Up and down the panel he went, until we heard a shout from Robert.

"The lights flickered," came the call.

That was enough for the professor. He yanked out that particular fuse and slipped a new one into its place.

"All right," shouted Robert.

A look of placidity returned to the professor's face. He carefully rebent the hair-pin into shape—and absent-mindedly shoved it into his waistcoat pocket.

CHAPTER V

# "IS THE DOCTOR IN?"

ALL kinds of visitors call upon me in the Park. Sometimes they simply bring me a story, as when the government man from the Canal Zone dropped in to tell me about the pleasant evening spent by two young men in Panama City, and the fight with the bushmaster which followed. There are my friends of the press and writers for the magazines. I have enjoyed devoting another chapter in this book to my pleasant contacts with them.

But many visitors arrive with pleas for help in solving their problems. Many of these problems are not associated with science, although it takes a scientist to solve them. There have been calls from the theatre for assistance in staging effects in which animals figure. I worked out one of these ideas in a way which was very effective.

The need in this instance was for the sound of howling wolves, which was to be heard during a backwoods scene. The producer was extremely temperamental, and brought with him to my office a man who claimed to be able to imitate any kind of animal. This man went through his repertoire, standing back from us at my doorsill. He cackled like geese, barked like a sea lion, roared, growled and chattered, to the astonishment of visitors on the exhibition floor. The place sounded like the Ark with discipline gone wrong. When he sought to imitate a pack of wolves I feared there might be a panic among the visiting public. It was this demonstration, for my approval or suggestions, that had brought the theatrical delegation to the Park. I suggested that as the wolves, which were supposed to be fraying the nerves of the isolated heroine in the cabin, were circling around at some distance outside, considerable soft pedal be used. The producer slapped me on the back and declared I had the very idea. We tried it again, but in softening the effect the imitator produced such mournful wails that my office sounded as if a wake were in progress.

I suggested that they bring a record cutting machine to the Park and make several phonograph records of our pack of wolves howling. The wolves howl at noon when the whistles blow. This was done, and later, on a stage steeped in blue light, and with a much distressed damsel wringing her hands and wondering if her rescuer would return with provisions to break a famine, the howling of those wolves gave the audience the creeps.

Another theatrical problem was more difficult, and remained unsolved. In this instance a cobra was to rear and bite a lady. The producer blew in, stated what he wanted, emphatically declared he had no idea of harboring a cobra, but wanted me to recommend some kind of snake that could be *dressed up* like a cobra, with something that looked like a flattened hood on its neck. His stage mechanic, it seemed, had designed the "hood" from an Indian picture. Now for some docile kind of snake that could be trained to wear it and rehearse for the part.

I told him that a snake could not be trained to rear like a cobra and pretend to bite a lady. We had a long argument, which culminated in my recommending the only good-sized serpent looking somewhat like a cobra which could be trusted. This was a Florida gopher snake. As the action supposedly took place somewhere on the Malay Peninsula, and black cobras are found there, the deception was not too far fetched. He was not impressed with my academic scruples, but I tried to be patient and gave him the better part of an hour.

I afterwards heard that they obtained a docile black gopher snake, which was dressed up in a flaring hood, tolerated its winged collar and appeared to do its level best, which was far from cobra-like, although the stricken lady helped by screaming loudly. I never had a chance to witness the serpent's acting as the critics turned thumbs down on the production and it quickly expired.

Several orchestras have come to the Park to try

the effect of their music upon the animals, the experiments being conducted at such places as the elephant enclosures, the lion or the monkey house, or in front of the series of bear dens. I have never noticed any other effect upon the animals than surprise at the outburst of sound. As to concentrating attention by fine blending and harmony of instruments, I think we could have obtained a keener reaction by having several of the keepers pound on some dishpans.

Requests for help in perplexing situations come frequently. One day a telephone call from the pier of a fruit company informed me that they were in trouble with a monkey brought north as a pet. It had escaped and evaded capture on the pier for a week. They wanted me to decide how it could be caught, and to start at the task immediately. I told them we were busy and would send down a man as soon as we could. This wouldn't do, they insisted, the man must come at once.

"Why the rush, about a monkey on your pier?" I asked them.

"Do you know that we have a row of electric lights more than thirty feet overhead?" came the reply.

"What about the lights?"

"That monk is unscrewing the globes and dropping them to hear them burst. It's a big job to run up the scaling ladders."

So here was another problem, not scientific but involving thought of how to trap a monkey on a five hundred foot pier with a roof trussed up by a maze of girders.

We rigged up a box on the pier and put a bunch of bananas in it. After a time the monkey slid down a girder and went into the box. We quickly shut the door.

Problems in the office? They are varied. During what I counted a fortunately uninterrupted morning I reached for a batch of technical notes and settled down to sort them. There was a knock at the door.

A keeper whom we were breaking in to assist the staff in the reptile house entered. He was a stocky Scotsman, who generally wore a good-natured smile; but the smile was gone.

"What's the matter?" I asked.

"A coppersnake bit me!" he declared.

"A copperhead? Show me the cage."

Within a minute I had a summary of what had happened. The offender was a Texas copperhead. The man had been cleaning the glass of the case and the snake had glided toward the front. The keeper sought to send it back to its corner with a wave of his hand, and the serpent had struck him on the index finger. Within a few minutes I had made drainage incisions and had a finger and forearm ligature attached. Then I applied the suction bulb. There was a breathing spell.

"What were you trying to do with that snake? You know it's poisonous."

"Sure, but who's afraid of a wurm like that?"

"Well, you see what you've got?"

"He's a peppery little devil, all right. Sure, and I feel sick. Me hand's burning."

Fordham Hospital is five minutes' drive from the Park. Here are skillful doctors, sterilized instruments, everything in readiness for an emergency. I pocketed a tube of serum, bundled the man into my car and started for the institution. The effect of the poison was rapid. Before we had gone a quarter of a mile the keeper was beaded with perspiration.

At the hospital, the serum was injected, the drainage incisions enlarged and suction treatment vigorously applied. It was decided to put the man to bed for a day, and to remove the wet dressing prepared for his hand at intervals so as to induce further drainage. I had no anxiety about his condition and was preparing to leave.

The man turned to me with a look of astonishment.

"And is that all I get?" he asked.

"Isn't that enough? What do you want the doctor to do? Chop off your finger?"

"No, and the doc has done enough, but haven't ye forgotten something yourself. How about the whiskey?"

I told him that it wasn't necessary in the modern,

scientific treatment of snake bite. His glare of disappointment indicated such disrespect for modern science that I was induced to summon a nurse. I considered the bulk of the man and his resistance to snake-poison, and whiskey. Whiskey does no good in a snake-bite case, and in moderate quantities does no harm. I indicated three generous fingers, in a standard tumbler. The eyes of the patient became beatific, his every-day smile returned. There might even have been a gleam of affection for the "wurm" that figured in the affair.

Another office incident was more startling. It opened with a call for an ambulance from our Small Mammal House.

"What's the matter?"

"Landsberg has chloroformed himself," was the startling answer from the junior keeper.

Landsberg—of all people!—one of the most competent and cheerful keepers.

The call went through for an ambulance and I rushed for the Small Mammal House. There I found Landsberg unconscious and heard and surmised enough to realize that something quite remarkable had happened.

A South American raccoon had in some way broken a leg. It was an old animal and possibly its bones were brittle. Landsberg had rushed to the veterinary department to summon Dr. Noback, chief of that department and in charge of pathological research. The doctor was in the middle of some microscopical work, and hearing Landsberg's story decided that the case of the old raccoon was hopeless. He suggested that the animal be mercifully chloroformed, and gave the keeper a partially filled pint can, also a big wad of gauze which he was to soak with the anesthetic and wrap around the wire front of the box in which the injured animal had been placed. Landsberg ran back, determined to end the suffering of the animal as quickly as possible.

He placed the cloth in a ball in order to saturate it immediately, then, imbued with vigor which should never be passed on to chloroform, shook the can violently as if it required thorough mixing. Under such agitation chloroform expands, and the contents of that can behaved true to type. The cork blew out and a torrent of the fluid shot over the cloth, but with such force that it splashed upward showering the man's eyes.

No one can imagine the agony produced by liquid chloroform coming in direct contact with the delicate conjunctiva, except by actual experience. In a frenzy of pain, which produced almost involuntary actions, the keeper grasped the cloth and pressed it to his face. He was gasping at the ideal rate to "go under" quickly, which he did, still clasping the cloth to his face. Fortunately, the junior keeper had seen the thing happen from the other end of the long building. He had hesitated for a couple of moments before grasping how serious the thing was, then rushed to his unconscious associate and torn the cloth away.

While the junior keeper was describing the accident the clanging of the amblance bell announced the arrival of a pulmotor. It wasn't needed. Landsberg was coming out of it. Some of the fluid remained in the can and I took a hand myself. As Landsberg saw the light, the raccoon peacefully passed out.

## CHAPTER VI

# DO SNAKES DIE IN THE SUN?

THIS chapter heading relates to a curious question I have heard on and off, along with the hoop snake story and the tale of the mother serpent swallowing her young. The latter two are old-established snake stories, and denial has brought sarcastic letters intimating that I was a desk-chair scientist, and didn't get around enough to see things happen. The first question has been put to me in correspondence, and when I casually denied that this was the case, the letter-writer seemed satisfied. The sun-loving serpent dying from exposure to the sun! The supposition seemed too foolish to deny, except in casual lines. I considered it one of the many snake myths. But after denying it I received a severe shock. Snakes *do* die in the sun. I had a clear demonstration of it. And afterwards checked and proved it. It occurred during an adventure resulting in the discovery of one of the most remarkable specimens I have ever seen.

It was hot, summer weather during a season when I had given up my annual vacation trip to complete some work in the motion picture laboratory. A letter had come from one of the investigators working on the idea of increasing the strength of serum for snake-bites. It dwelt upon my studies pointing to the higher toxicity of poison in snakes coming from the hibernating dens in the early spring. In such specimens the poison has remained in the glands without any expenditure or renewal for about six months. During this long period the temporal glands, which function to change the fluid which forms within them to a lethal poison, has distilled it to the highest power. I had sent poison from freshly captured "spring snakes" to the laboratory, the venom of the mountain rattler. The writer of the letter explained that certain proteins had been noted in that poison, which appeared to be lacking in other poisons "milked" from snakes in the laboratory. Captivity might have affected the latter specimens. The laboratory wanted some poison from freshly captured "summer snakes," which in hunting about and biting their food were at rather frequent intervals expending poison, and so in a condition when their poison glands were being partially emptied and refilling. Were these proteins lacking in the wild "summer snakes"? The idea was to obtain the most powerful poisons to immunize horses and hence obtain a stronger neutralizing serum. If the "spring snake" was found to secrete a venom with toxic elements superior to, or almost lacking in the "summer snake," the use of its poison would produce a more efficacious serum.

I was glad to take the matter up. The work in the laboratory had been running into monotonous channels and a change would do me good. To help the toxicologists it would be necessary to go up to the Berkshires, the only mountains within a hundred miles of New York where there is a fair chance of finding rattlers during the summer.

The reader may wonder why the reptile expert can't find rattlesnakes at any time of the year between spring and autumn. There is a big difference in the hunting for snakes according to the seasons.

In the spring the reptiles are emerging from their hibernating quarters, to which they have come during early autumn from distances of several miles around. They lurk near the rocky dens for several weeks, taking no chances of being caught far away from deep, sheltering crevices during the fickle weather of April and May. With the latter part of May, they are on their way to surrounding woods, meadows, stone walls, and rocky canyons, some of them appearing to bear in mind the ruins of old buildings with crumbling foundations, piles of old boards or sawdust mounds. Snakes, like birds, after leaving places where they have spent the winter return to favorite spots where food is plentiful. The difference is that birds fly to warmer climates and wing their way back over hundreds or thousands of miles to the temperate latitudes. Snakes go into winter quarters, deep into the ground or rocks to keep from freezing, and glide back through a labyrinth of declivities to their summer hunting grounds. To me, their sense of location in returning to the dens, and on leaving these shelters scattering to their favorite places for the summer, is as remarkable as the migrations of birds. It points to the same mysterious sense of direction.

By the time summer is well established, the occupants of the dens have gone radiating out in all directions. They are widely scattered, and the problem is to find the spots to which the individuals have gone. The reptile expert has an idea of the kinds of places they may select, but what looks good to him, may not be favorable for the snakes. Small rodents may be lacking. The area may be too damp, or too dry, particularly the latter. Rattlesnakes are not to be found in a spot where water to drink is inaccessible. Yet the collector may be deluded and pass a good place, not visually indicated as worth careful searching. No water is in sight; yet a hidden spring under a jumble of rocks may make it ideal for the snakes.

Why do they leave the mountain side and the dens in the first instance? Because most of the dens face south and the rocks are hot and dry during the summer. There is no access to water. Then again, a large number of reptiles in one place would quickly clean up all the food.

In the summer I have investigated miles of country

surrounding populous spring rattlesnake dens, and during weeks have not seen a snake. It should be carried in mind, of course, that after feeding these creatures hide; and even though coiled outside, it is easy to pass the mottled, camouflaging pattern of a rattler, provided it lies quiet. That they are here and there is clearly demonstrated. The country people in their trips for berries encounter them, but to hunt for rattlers in summer is doubtful work.

This was the prospect after reading the letter from the research worker requesting fresh poison. But I knew of a spot where rattlers lurked all summer, not far from an extensive den. It was a mountain top, with shelving rocks and knee-high huckleberry bushes. Birds came to the bushes for berries. The rattlers knew it; and between hunting and feeding, the big flat rocks afforded good hiding places.

That spot in the Berkshires is known as Black Rock. It rises on the westerly side of the valley in which nestles the quiet little town of Sheffield, with its elm-arched main street. Throughout this valley upon which the mountain looks down, forming a sort of massive step against the far higher and impressive bulk of Mount Washington, are farms of meticulous neatness. It is the country so well described by Oliver Wendell Holmes, particularly in his strangest novel, "Elsie Venner," in which the rattlesnake theme, sinister, half hidden, runs throughout. Curiously enough, rattlesnakes are seldom seen on the farmlands. They keep to the mountains, and there are common enough. Yet strangely, too, accidents in the Berkshires are almost unheard of. In twenty years of visits to that country I have heard of but three cases, each due to unusual carelessness in places where rattlers were known to lurk.

Black Rock looks as if it were close to the Under Mountain Road. Its name comes from a wall of blackish rock several hundred feet in height near the top of the mountain. From the motor thoroughfare it looks smooth and vertical. Its base is smothered in trees.

There is a considerable hike from the road, however, to reach the foot of the mountain, and when one arrives at the base of the rock itself, there is a devil's den of huge fragments of broken stone scattered pell mell. Among these masses I had previously seen occasional rattlers in summer, but on this trip I pushed on to ascend the mountain. At one side of the precipitous face, invisible from the road, is a cleft. It is so narrow in parts that it is necessary to squeeze through and upward. With soft-soled shoes the ascent is not dangerous, for roots project here and there and may be grasped; but each foothold must be studied. Descent is more difficult, particularly when one is swinging down a bag of rattlers. This was my route, the first resting place being the top of the rock.

Here the view is magnificent, sweeping out for

miles southward and over the lower Taconic Hills. After resting for a few moments, I pushed on. The remaining climb to the summit is more gradual, though through one of the worst tangles of scrub oak I know of anywhere. The summit of the mountain is quite flat, a mixture of wiry little oaks, huckleberry bushes and partially imbedded shelving rocks. Under every one of them there are crevices where it is possible to run a thin pole in six feet or more to where the opening turns and goes deeper. There probably isn't a rock on that mountain top that hasn't sheltered a rattlesnake. I have probed those crevices many times and heard the warning buzz from within. Sometimes it is possible to so poke the rattler that it will come raging out.

I had sought to pick up Harold Roys, who lives near the base of the mountain, but he was away. Harold knows every likely crevice on that mountain and for years has gone up and studied the rattlers. I spent two hours writhing through scrub oaks and peering into crevices, for rattlers lie in summer under the edges of the rocks. The temperature was high, about eighty in the shade, and my shirt was soon soaked with perspiration. It seemed not to be a day for rattlers, and I was about to give it up. It is often this way. You can search for hours in a likely place where you know there are snakes, and nothing is out.

Descending some step-like projections I came on two rattlers, in symmetrical coils which almost touched.

One was so black it looked like a pad of velvet. The other was rusty brown. That is the way with the rattlers of Black Rock. The clan exhibits a variety of colors, from roll sulphur to pitch black. The snakes were under the low, spreading branches of a dwarf oak. There was a crevice a yard away. They were watching me, and alert, as I could see from their darting tongues; but they refrained from rattling, having the idea that I had not seen them and would pass by—a common trait with the mountain rattler. I knew they would rush for the crevice unless I took lively measures, so I seized my stick, swept one of them out to an opening, and went for the other, which was gliding for the crevice. It was a case of handling both of them. There was no time for noosing. I pressed down the head of the brown one, got him by the neck and fought him into the bag. Giving the bag a spin to close its upper part I swung around for the other. His rattle was now going full tilt, but he was also making for a crevice. He went into the bag.

This was gratifying. Two rattlers on the first day, and there would be more hunting. Continuing the descent, I saw a big rattlesnake going under a rock below me.

Dropping the bag, which I had tied shut with noosing cord I slid down after him. He was in a mean place. It was necessary to work around a shoulder on a ledge where the rock sheered off precipitously. Far below, where the mountain sloped outward, trees looked no larger than bushes. Getting a foothold and peering into the crevice where the snake had gone I could see him peering at me. He seemed to feel secure for he did not rattle. There was only one way to get him, and that was to fasten a noose on the end of my stick and snare his head. There was not much space, the crevice being barely two inches high, but there was a chance. Twice, I had the noose under his chin, but his head was high and touched the top of the rock. Then, assuming the attitude that he had sparred with me long enough, he turned and glided into darkness. I had spent about fifteen minutes in the effort to get him.

Returning to the bag I picked it up and resumed my way down the mountain. Something about the feel of it seemed strange, but I was not actively aware that anything had gone wrong until I started to slip and slide in getting down some steeper places. Rattlers coil and become quiet in a bag, but these felt soggy. I opened the bag and looked in. From the limp loops, half on their sides, I saw that both rattlers were dead. They had been in no way injured during capture. The bag had been laid on a sunny rock, placed there without looking where I put it, while I had kept my eyes on the spot where the third rattler was going in. Here, I confess, after years of association with serpents, was a humiliating discovery. Snakes died in the sun! The summer sun—and within a few feet of where they coiled a few moments before! Later, I checked the matter to satisfy myself, and later still, as I will show, it was made the subject of detailed scientific study.

So my first day's hunt on Black Rock was a failure. But there were other trips, and several rattlers were captured. On the last day that I could stay Rolls Smith, of Sheffield, went up the mountain with me. He is a keen naturalist and would rather hunt rattlers than fish for trout.

There had been a shower the previous evening and the morning dawned with a cool breeze. I figured it to be the kind of day that would lure rattlers from their hiding places—and it was. We captured four big fellows. Some smaller ones glided for crevices, but I didn't bother them. Laden with the bag we descended the precipitous crag. In some stretches we let the bag down ahead of us by a stretch of the heavy noosing cord, so that it could rest on a projection, and thus free us to use both hands in grasping roots or niches in the rock. At the base of the great wall we followed a tortuous downward course through a labyrinth of shattered stone. On a table-like rock below was a white object.

"What's that on the rock?" I queried.

Rolls Smith looked down.

"Somebody has been up here painting, I guess."

Artists occasionally climbed as high as this on the mountain. The slope of the great ravine, looking toward Mount Washington, is well-known to the oil-painting colony, which annually comes to the Berkshires.

Mild curiosity prompted our working towards the object to see what it was, but coming to its level I had to get past a chunk of rock that towered over my head, then move around treading on fragments that teetered and called for careful footing. I was more intently watching where I was placing my feet than glancing ahead at the objective, but eventually I felt myself on firm ground. The table rock was now only six feet ahead. Glancing forward I saw something that made me gasp. It was a coiled rattlesnake, as white as alabaster, with glowing, pink eyes.

There was a touch on my shoulder. Smith had worked around the big boulder to share the small area of firm footing.

"Look at that!" was my exclamation.

With the words, I unwisely extended a pointing finger. The action was involuntary, emphasizing the strangeness of such a sight.

"Well, what do you think—"

Our spellbound attitude consumed a fateful half minute and my pointing hand was a badly timed gesture. The white coils swelled with the intake of breath; there was the sharp buzz of a rattle. Almost simultaneously all parts of the creature started to move, revolving, and from the circle came the head and neck. As we stumbled over more teetering stones to gain the table rock, the reptile straightened, glided over the edge, turned, and as quickly disappeared under the big rock. The buzz of its rattle grew fainter, but continued.

Investigating the spot where it had gone in, we saw that the table rock rested on a bed of stone. The under-surface of the flat shell, which weighed tons, was slightly concave. This produced the sheltering crevice which, without doubt, led a considerable distance inward, and probably downward—for rattlesnakes are wise enough to select secure retreats. We could still hear the rattle buzzing faintly, then the sound faded. That snake had gone to a depth from where there was no hope of retrieving it.

It is said that albinos have poor sight, but that white rattler had behaved as if its vision was as keen as any of the normal residents of the ledge. I figured that it had a keen sense of danger, as albinos form a glaring target for enemies. That probably was the reason for its living in the secluded labyrinth at the base of the ledge. Before leaving the place I looked at the spot where the prize had been coiled. There was a bed of mottled brown leaves, with some tufts of feathery green lichen.

On the way back to Sheffield, Rolls Smith said that he would get over to Black Rock and watch for the albino rattler. This was comforting.

Several weeks later he wrote me that he had been to

A LEDGY SHELF ON BLACKROCK

Near this spot the author noted that "snakes die in the sun."

THE WHITE RATTLESNAKE

One of the strangest discoveries in years of collecting. The spots along the back were of pale yellow.

*Permission of The Daily News*

A FAMILY OF GOLDEN MARMOSETS

There are twin babies, one being partially hidden in the background. Strangely enough the male took charge of the infants, carrying them on his back, except in regularly returning them to the mother for short periods of nursing.

Black Rock several times, but had seen no sign of the albino. Time went by and in the parks around New York there was a red tinge among the trees. I thought of the white rattler and formed a mind picture of the Berkshires, now beginning to blaze with color—varied red on the mountain sides and yellow through the ravine leading to Black Rock, as the broad leaves of the wild grape changed color. The picture was alluring, and I made up my mind to go up again. While waiting for a chance to get away, I came across a letter from Smith in the morning's mail. I opened it casually, expecting a chatty note, and assurance that he had done his best in searching for the strange specimen.

The sheet contained a mere note and appeared to have been written in a hurry. It read: "Have caught the albino rattler and am boxing and shipping it this morning. Wire me if it arrives all right. Coiled in the same place, but nearly fooled me—so many leaves are down."

The specimen arrived in perfect condition. It attracted wide attention and became quite tolerant to being "posed" for photographers after illustrations for newspaper and magazine articles. Among the photographs illustrating this book there is a picture of it.

The curious sequel to the story is this: A year later Smith was on a stroll through the ravine and decided to return along the base of Black Rock. He was carrying a snake bag. Leisurely poking and peering here and there, as a naturalist does, he came upon *another* albino rattler, close to the table rock. Having already seen one albino on the mountain, he was consequently proof against the "buck fever" which I had exhibited. He captured the serpent and shipped it to me.

A comparison of the number of segments of its rattles with the number possessed by the examples already at the Park, led me to believe that it was of the same age. It was possible that a litter of albino rattlers had been born. Looking over my notes relative to the number of young in litters of timber rattlers born in captivity showed me the following counts: 12, 9, 12, 7, 9. Possibly there were more albinos on Black Rock. The next spring I spent a week investigating those ledges. There were plenty of ordinary rattlers, but no albinos to be seen. Rolls Smith and the veteran rattlesnake observer Harold Roys have since searched and failed to discover any.

But to return to my accidental discovery that snakes die in the sun. The matter was taken up systematically by Dr. Walter Mosauer of the University of California. It was he who surprised veteran reptile experts by discovering how a serpent can rapidly glide by anchoring one or two of its undulated loops against a projection of the ground and "feeding" its body ahead of the anchorage. The doctor proved his contention by placing an active snake on a polished surface, where it "swam" in an effort to go forward. The polished surface was provided with a few holes. When the scientist stuck some short pegs in the holes and the snake obtained a body purchase, it glided off at a rapid rate. He provided additional proof of his theory by analyzing the actions of the snake in slowed-down motion pictures, both with the pegs and on surfaces of soil and sand.

So this scientist went after the solution of snakes dying in the sun. For the purpose he selected the desert sidewinder rattlesnake, a small rattler which receives its name from crawling in laterally thrown loops to keep from sinking in the soft sands of the sterile wastes of our Southwest. Sidewinders prowl the deserts at night, and the specimens used were captured after dark with flashlights.

The experimental specimens were placed out in the desert sunlight, where the air temperature was 35.5 degrees Centigrade (about 95 degrees Fahrenheit) and the temperature of the sand was 55.5 degrees Centigrade (about 130 degrees Fahrenheit). The first snake died in seven and a half minutes. The second died in nine minutes. A desert diamond-back, a much larger snake, died in about six minutes. The temperatures of the snakes were taken and they were approximately 47 degrees Centigrade (about 115 degrees Fahrenheit) at death.

In an article * written in collaboration, Dr. Mosauer

*Copeia, No. 3, 1933.

and Dr. Edgar L. Lazier, also of the University of California, say:

"The most significant observation seems to be the almost identical body temperature in all specimens just after death. Apparently the snake's body cannot withstand a temperature above 46 Centigrade (about 115 Fahrenheit) and the death is caused by overheating."

Lloyd W. Swift, of the United States Forest Service, sent an observation to the same scientific magazine which published Dr. Mosauer's article. He drove a rattlesnake into a sunny gravel pit and it died in twenty minutes. Mr. Smith's note reads:

"A Pacific rattlesnake (*Crotalus c. oregonus* Holbrook) was killed by exposure to full sunlight in twenty minutes on August 15, 1932, near Bucks Lake in Plumas County, California. The altitude there is 5000 feet, the exposure southwest, and the day was clear, quiet and warm. The snake was fourteen inches long and had four rattles."

Since these observations were made it has been noted that other serpents besides rattlers die in the sun.

To appreciate the significance of these studies it should be understood that the temperature of a snake's body is entirely influenced by the temperature of the surrounding air. At 70 degrees Fahrenheit, a serpent is normally active. It will seldom feed at a temperature as low as 60 degrees. At 50 its actions are slowed down and it appears sluggish. If the surrounding air is sufficiently cold to reduce the temperature of its body to 40, it may appear as limp and devoid of motion as if dead. Most snakes are killed by exposure to 32 degrees, or freezing point. That is why they retire into deep crevices or dens for the winter, seeking depths beyond the possibility of penetration by frost.

If the environmental air (as the scientists say) is stepped up to 80, 90, or even 100 degrees, the snake's body temperature likewise rises. The observations demonstrate, therefore, that the snake can continue to survive temperatures up to a point when its susceptible body is heated to 115 degrees, and then it dies. Hence the prowling snake in summer must use caution about gliding in the sun, and without doubt its sensitive, scaly surface gives it ample warning when it is traversing areas of ground where considerable variation in temperature is produced by vegetation or dampness. The sun of the spring and fall is relished by snakes, which openly bask in it for hours.

This all goes to show that there may be a foundation for the so-called myths and superstitions. Do snakes die in the sun? It sounded like a foolish question. I still deny the hoop-snake story and politely seek to disprove the allegation that the mother snake swallows her young for protection. Politely, I say, because that is a favorite tale, and so many people firmly believe in it.

CHAPTER VII

# THE JUNGLE CIRCUS

My movie laboratory has become and continues to be an important part of my work. My idea when I built the studio was to prepare a living book of natural history. At the time of my trip to Black Rock I had proceeded through mammals and reptiles and had reached insects. I am going to make some confessions about my work with the motion picture camera, and tell about some developments which shocked me as a scientist.

My assistant was a young electrician who aspired to work into the technique of the movies. Andy was slight and freckled, and his hair had a tendency to stand on end—probably from a habit he had of running his fingers through it. This came from thinking. He was never satisfied with anything—cameras, lights, switchboard, transformers or projecting machine. He was constantly making alterations, but I trusted this young wizard, who seemed steadily to make everything better. True enough, my heart sank one evening when I found he had the delicate mechanism of my camera apart.

"It's binding in the gate," declared Andy. "That roughens the emulsion and gives you a rub when the film goes into the valve of the take-up magazine. That's why you're getting static marks on your film."

The reassembled camera functioned perfectly, and there was no more static.

I had been making motion picture records of insects and allied creatures. Several thousand feet of film recording the habits of the local spiders had been filed. Those were fascinating scenes. They related to the care of the eggs and the early attention to the young. Spiders show the most scrupulously systematic habits along these lines. There was one large local species, of the genus *Dolomedes,* which had taken a lot of time. It is a water spider, that is to say, it lives in damp places and can run nimbly over the surface of the water owing to velvety hairs beneath the limbs, which carry a veneer of air. If one looks upward through the side of a glass tank to a specimen of *Dolomedes* resting on the surface, the undersurface of each limb looks as if it were coated with quicksilver. I had collected half a dozen female specimens, which were carrying their egg cocoons.

With *Dolomedes,* the eggs, like minute lustrous pearls, several hundred of them, are laid in a little silken net, which the mother spider gathers around and seals into a bag with silk from her spinnerets. She twirls the mass, further swabbing it until it looks like a silken ball close to half an inch in diameter. This she carries beneath her body, gripped with the sharp points of her fangs. If she drops it for a moment to consume insect food, she stands close by and again seizes it upon the slightest disturbance. She can be torn limb from limb rather than relinquish the precious burden.

My female examples of *Dolomedes* were kept in separate trays, covered with a panel of loose glass. When I wished to photograph them, I placed the tray under the camera for a vertical shot. Focusing was done through the glass. When all was ready the glass was lifted off. The spider usually remained in the tray. One would occasionally dart out. I was prepared for this with a silken net having a ring no larger than a soup spoon.

These females were photographed to show how they carried the silken bags of eggs, and how they fought if an attempt was made to take the eggs away from them. I had designed a pair of snub-nosed tweezers to grasp and "borrow" the egg case from a female spider, without injuring her. The operation was tricky, for it was necessary to encase her in the little silken net as the eggs were taken or she would come chasing after them. In this way ultra close-ups of the egg cases were recorded, and in one instance the case was opened to show the gleaming mass within. When the injured bag was returned, the spider carried it as before. In another instance a bag of later development was opened. It was figured that the little spiders had hatched inside it, and this was the case. The legion of infants came trooping forth. They were a gleaming, golden yellow. The field of photography was not larger than a post card. When the scene was projected the little spiders looked like clustered blossoms in a daisy field.

When the mother was liberated from the little silken net in the middle of her infant flock, she immediately started the construction of what I called the nursery web. This is the only time that the prowling *Dolomedes* spins a web, while the young are gaining strength after their first emergence to sunshine and air. Here she stands on guard fighting off intruders, while the young absorb the egg yolk, after which they are ready to scatter and fend for themselves. Normally, the female tears open the egg bag with her fangs and liberates her family. I watched for this, but failed to see it. In each instance it was done during the night and I found the nursery established during the morning's inspection. The nursery, however, and the activities of the mother were duly recorded.

Then there were the habits of the mother *Lycosa*, the wolf spider of our northern fields, another stalking species which spins no web. *Lycosa* is the largest of the local spiders, rather stout-bodied and of a dark, velvety brown. There is the same construction of a silken case for the eggs, but with the additional spinning of a strap across it and the fastening of the ends to the body. Thus *Lycosa* is not so encumbered as *Dolomedes*, and can run rapidly with her fastened burden. This attachment kept me from making close-ups of the egg-sac of this species. I didn't feel at liberty to cut away the silken harness. *Lycosa* also rips the case open with the fangs, and thus, by watching a series of trays, almost living over them at the critical time, I was able to film the emergence of the infants. Then the strangest of sights occurred. The young spiders climbed upon their mother's back. There were so many that they formed a double layer; but they adhered so tenaciously and so compactly in a clinging mass, that an uninitiated observer, glancing at the parent, might have remarked about the pebbled condition of the big spider without noticing anything unusual. The circumference of the spider's body, however, had now been increased fully a half by the attached progeny. With a fine camel-hair brush the little spiders were rubbed from the mother's back. They rushed and milled about her in eddying circles, while she endeavored to fight the brush. When the operations with the brush ceased, they utilized her limbs as gangways and were soon aboard again. Thus *Lycosa* carries her young until they are able to look out for themselves.

A series of high cases, with glass fronts and backs painted dead black, was reserved to record the exquisite work of web-spinning spiders. Our work with these was done under very different conditions. The wild spiders were sought in the early morning, when dew on their webs made them easier to find. My little silken net was used to transfer the spinners to small cylinders of wire window screen. Each spider was placed in a separate glass-fronted case, where it would crawl into a corner and remain quiet all day. Several branches in the case formed an inducement to construct a web.

The orb weavers spin their complex structures at night. A web may last a long time if the weather is calm, but if it is destroyed by rain or wind, a new web is spun the following night. So we watched our freshly caught web spinners. They usually had new webs under way around ten o'clock. I would select the beginning of a web which indicated that the structure would run parallel to the front of the case. Then the glass would be slipped out and the studio lights turned on. These operations disturbed the spider, which would retreat to one of the uprights and sometimes cause a delay of half an hour before resuming operations. The placing of the spokes of the web was a slow process, as in many instances the worker had to use an upper spoke to swing along in order to attach the next lower one. The spiral portion was quickly made, the spider walking about in rapid circles and attaching the taut, dragging thread to each spoke with a single move of her body. Some of the webs were completed in two hours; others took considerably longer. The lighting had to be arranged to make the silken threads gleam and show against the black background. A powerful light through a lens was shot in from the side to do this.

Two types of large local spiders were the star web-spinning actors, *Miranda* and *Metargiope.* They construct the most symmetrical webs of our northern spiders, and any observer afield can see such webs in the late summer among tall stalks of goldenrod.

With the spinning recorded, the spiders were carefully fed day by day as I waited for another phase of construction work, the spinning of silken egg cases of far more complex structure than the bags carried by *Dolomedes* and *Lycosa.* First there is spun a silken bag, then around it a thick fluff of shining silk. Over this is spun a tough and waterproof covering. The final structure is of astonishing size when compared to the builder. With both kinds the casing is the size of a fair-sized marble. With *Miranda* it is swung in a strong silken cradle, among tough, high-standing stalks close to the web. With *Metargiope* the casing is like an inverted balloon in form. These casings are waterproof and winter proof. The eggs do not hatch until the following spring and there is no mother there to tear an opening through which the young may emerge. Concerted effort supplementing a silk disintegrating fluid from the mouthparts allows the little spiders to make their way through. The mother has died the preceding autumn, with the first frost. The young attain maturity the year they hatch, and similarly perish with the first frost of the same year.

Thus the stories of interesting creatures accumulated on the film.

Along with the insect work, I was developing films showing the types of locomotion among different kinds of small creatures. There was the gait of a desert tortoise, moving on feet in miniature like those of an elephant, slow, but as steady as a mechanical contrivance. Another illustration was the tree toad, climbing a smooth, vertical surface with adhesive finger disks, and balancing or working around a slender twig like an acrobat on a horizontal bar. There was the rush of several kinds of lizards, one of which reared upon its hind legs by balancing with upcurved tail. There was the gripping of branches by the African chameleon, with its bifurcated feet. The chameleon occupied the stage for some time. I recorded in close-up detail the independent rolling of its goggle eyes. One could look forward and the other upward, backward and downward. Here was specialized development allowing the chameleon to squint forward while taking aim at a fly with its long, sticky tongue, and simultaneously roll the other eye backward on the lookout for intruders. Andy and I had some good laughs when this film was projected on the studio screen.

But a serious condition was developing. My money

was giving out. Film was expensive. There was the cost of the negative, the developing, and the additional expense of positive film for projection. From the start I had made up my mind that I wouldn't borrow any money or ask for help. I wanted to do exactly as I planned and pleased in building up a series of films, without accounting or outside advice. The winter lectures had supported the work, but the funds from that source were gone. There was only one thing to do, and that was to sell some film material. I made up several reels and took them to one of the motion picture companies.

They said the scenes were fine, but they didn't like the arrangement. They suggested a continuity of the scenes which clashed with my ideas of describing the subjects. Another company had a similar reaction, but suggested an entirely different arrangement, with "funny" titles. It was the same story, all along the line. Everybody had a different suggestion. I came back with the reels, much dejected.

Andy and I had a conference. He insisted that there wasn't any need of my keeping up his modest salary. That could stop until I was on my feet again. But I told him that while I appreciated this spirit, we would have to get money to carry on normally, as there was the film to buy, and electric bills and installments on some of the lighting apparatus to pay. I must seek a solution to the arrangement of my film scenes.

Andy had browsed around enough to know something of the film business. He said that if there was any place in the world where suggestions were in the air, it was among the movie people. They spent a great part of their time suggesting changes in each other's ideas. If they couldn't see a picture without making suggestions, they figured the time wasted. Andy insisted that many of the executive officers of the big companies spent more time in giving advice in projecting rooms than in doing real work. He mentioned an instance where a man directing a picture knew what he wanted, and finally got it by working along unusual lines. There were but two ways to make the picture, and one was the way that the director knew was wrong. So he started in that way, pieced up the subject for experimental screening, and had it run before the conference board. Suggestions were emphatic that it should be arranged the other way—which was the way the director wanted. So as to make sure of his ground, he hotly debated the changes, was told that the suggestions must be carried out, and chuckling to himself completed the picture according to his secret desires.

"But I can't do anything like that with these fellows," I told Andy. "How am I going to satisfy them and sell some pictures?"

Andy waved a cigarette in defiance of a "No Smoking" sign. He launched into a torrent of suggestions more discouraging than those I had heard during the peddling of my reels.

"I'm looking for practical advice!" I exploded. "You're worse than—"

Andy extended his cigarette in a restraining gesture.

"The trouble is, doc," he said quietly, "you're a scientist."

"And who could make those pictures who wasn't a scientist?"

"Right," said Andy. "You've got some swell stuff, but the guys who run the theatres want to hear some laughs from the audience."

"I'm not making comedies."

"If you'd hook up some stuff to get a laugh, and shoot in some snappy titles, you'd sell the film—if you gave those guys a chance to make a few changes—just to satisfy them."

My dignity as a scientist received a jolt. My fine material arranged as comedies! But as Andy's advice continued, the jolts came harder.

"There's the cock-eyed lizard," he suggested. "He'd get a big laugh."

I winced, as my studies of the independent eye movements of the chameleon, were thus designated.

"There's the frogs with the suckers on their feet," continued Andy. "Work a couple of 'em on a horizontal rod. Twist it around and if one falls off, all the better. There's a subtitle for you—"The On-and-Off-Again Brothers."

By this time my forehead was moist.

"There's that paddle-foot frog that never comes out of the water, swims on his back, and does some grand stunts. Call him the Diving Venus."

And this related to my scenes of the strictly aquatic *Xenopus*, from Africa!

I was stunned by the suggestions. They were like darts puncturing my scientific prestige. I glared at Andy, but he wasn't through.

"You've got enough stuff with laughs to make up about four episode reels and call 'em the Jungle Circus series. There you are, doc, you can knock their eye out with a set of reels like that."

"And lose my reputation in doing it!"

Andy regarded me tolerantly.

"If any of the highbrows come back at you," he said, "tell 'em you made the reels to get the kids interested in animals. You've got to sell some film, haven't you? You can see the highbrow stuff doesn't go, can't you?"

With a guilty feeling slightly subdued by the thought that such pictures would interest children, I started the preparation of the Jungle Circus series. Andy's advice continued to be helpful, but nevertheless shocking.

The first thing we did was to make moulds and cast a series of toadstools in plaster. When a group had been set up on a bank we collected a batch of toads, and to calm their restlessness put them in the refrigerator for a quarter of an hour. They were chilled just enough to sit passively on the toadstools, all facing the camera. Coming at the beginning of the reel they showed that a portion of the "audience" was assembling to see the jungle circus. A scene of the chameleon followed, and a title indicated that he also was a member of the audience, who fortunately could gaze upon aerial acts with one eye and down into the arena with the other at the same time. Various odd, small creatures appeared in flashes, as members of the gathering clan.

To open the "acts" of the jungle circus, we selected a series of flying leaps from the end of a branch by some little animals known as galagos, lowly members of the monkey tribe, inhabiting Africa. This was followed by some hurdle jumps by African desert jerboas, which are long-legged kangaroo rats. I had already filmed them leaping over fair-sized stones, but Andy insisted they should be "worked" again jumping a row of professional-looking barriers. The effect was astonishing. They went over the obstructions like trained kangaroos, as if they had been rehearsed to do it. Andy beamed.

The acrobatic tree toads followed and we found that they could climb a thick cord like gymnasts going hand-over-hand up a rope. Their expanded and adhesive toe-disks enabled them to do it. There were scenes of them going up the "ropes" to their aerial apparatus. For this we tried some experiments, one with a turning cord, another with a turning rod. The effect was of acrobats doing the giant swing, except that the dexterous forefoot grasps and changes made in adapting the performers to difficult positions exceeded human capabilities. Occasionally one of them would fall off, but we had a soft layer of cloth underneath and out of sight, and the performer was not injured. We ushered it into the scene again by sending it up the rope to the horizontal bar. Truly these tree toads deserved the title of the On-and-Off-Again Brothers. Their antics were ludicrous and we had some hearty laughs ourselves.

The "big scene" in this first reel was one of my own design and related to the performance of a fly—a bluebottle fly.

During my studies in entomology I had made some observations relating to the strength of insects, and one of these had been the pull of a fly's feet when under restraint. Flies, like tree toads, have adhesive pads on the feet. They are so small, however, that they cannot be seen except with a magnifying glass. It is the action of these pads that enables a fly to walk up a vertical surface of glass, or to walk across a ceiling. In proportion to their size the strength of insects, as compared to that of four-legged animals, is enormous.

Remembering my former experiments in testing insect strength, it was my idea to film a scene of a fly juggling a dumbbell huge in proportion to the performer's size—a typical strongman act. The idea of that particular stunt was not new. I had seen it done in England, for the amusement of a group of children, but I had not seen it reproduced, close-up, on motion picture film. However, I decided to give it some embellishments to produce a background to fit the reel.

A dumbbell was made by trimming down a couple of small corks until I had two balls, each a quarter of an inch in diameter. This was delicate work. The balls had to be smoothed and rounded with fine sandpaper until their outlines were perfect. I had to reckon on the great magnification of the projected picture. The stem of the dumbbell consisted of a fine hollow straw.

A miniature chair was constructed. It was half an inch high and this was cemented to the domed back of a big, tropical beetle, with an elongated and down-pointing head. This lumbering beetle was close to the size of a bantam's egg, and had a habit of walking a few steps, stopping for a minute, then going on again. With the chair on its back it looked like a miniature elephant ready to bring a star performer into the ring. The next job was to place the fly in the chair, and keep it there.

This was done by putting two minute specks of gum against the back of the chair. Several bluebottles had been enticed into a flytrap by baiting it with stale meat. Selecting one of them I held it gently between my fingers while Andy pressed its wings against the back of the chair, where the gum held them in restraint. The fly could do nothing but wiggle its feet. The position made it appear as if it were actually seated in the chair, in a semi-reclining position.

The camera was ready and focused on a flat area designed to represent the floor of the arena. The beetle was taken back outside the camera lines to a distance calculated to have it walk far enough, and make its periodical stop, in front of the camera. The dumbbell was presented to the fly.

The idea of the induced performance was simple. The fly objecting to the restraint waved its legs in seeking something to grasp by which it might pull itself away. The dumbbell seemed to be something that would help. Grasping a spherical end, the fly pulled with a treading motion of its feet. The dumbbell not only revolved like a spindle, but the treading motion swung the whole affair until the ball at the other end turned in an arc toward the fly, which transferred its grip in seeking a point for a strong pull. The effect was of a rapid and remarkable juggling of the dumbbell.

The fly was performing in this way when the beetle was liberated and started toward the camera. The fly's performance was marvelous, but during several attempts the beetle stopped in the wrong place. Andy's forehead was beaded with perspiration. He was using explosive language, directed at the beetle. The film gauge on the camera showed a waste for each unsuccessful run, but at about the fifth attempt the beetle stopped almost in the center of the arena, and the fly's performance was all that could be desired. The camera purred through a long and excitedly satisfactory scene, and then, to our gasp of joy, the beetle reared on its stubby legs and carried the performer out of the scene. That perfect though accidentally timed exit came at a point when the gauge showed my film magazine pretty nearly exhausted.

As I gently liberated the fly's wings from the dots of sticky gum, opened my fingers, and the insect streaked away, Andy glared at me in consternation.

"He was a knock-out!" he declared. "Suppose the film isn't all right."

"They all work the same way," I insisted.

Andy was not convinced. His attitude indicated that he blamed me for losing contact with a particularly competent fly.

When the film was developed the scene was found to be perfect and thus the first reel of the jungle circus was completed.

The "big scene," as Andy called it, in the second reel was made by selecting film already in the files, and also was my idea, but prompted by Andy's original suggestions to "put punch" into academic material. The idea was to produce an eating race, the contestants to be a grasshopper, a caterpillar and a monkey.

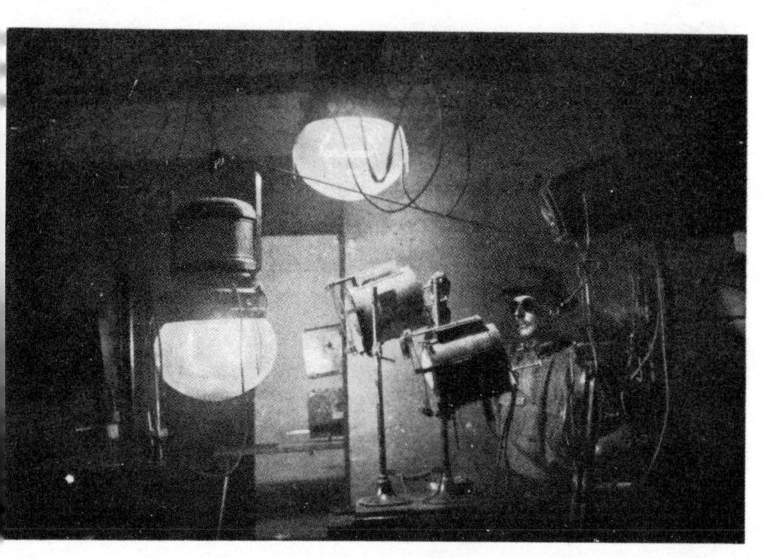

IN THE SCIENTIFIC MOVIE STUDIO

All the complications of the dramatic plant are here, and the actors are uniformly temperamental, some refusing to "work" except at night.

A STAR ACTOR

The African chameleon, with its droll actions and independent eye movements, solved a tragic development in the studio.

TREE FROG

In company with the chameleon (seen on accompanying plate) this specimen played an important part in an odd period of the movie studio's history.

LIFE HISTORY OF THE FROG

The scene indicates one of the many phases of study possible in the scientific motion picture laboratory.

As the grasshopper and caterpillar had been filmed close-up, their screen images were as big as the monkey's. The grasshopper quickly consumed a stalk as long as himself; the caterpillar, by rapid, scissor-like movements of its jaws entirely devoured a leaf, also as long as its body; and the little monkey cracked the shell of an egg and lapped out its contents, making a mess of the job toward the end.

By mixing up the scenes, after introducing the three alleged contestants in rotation, the continuity became really exciting. When the grasshopper was half through his stalk, grasping it between forefeet and chewing vigorously, the monkey was flashed in, with the egg-shell bitten away to the shape of a cup. It looked like a neck and neck contest between these two until the caterpillar entered the picture, the leaf more than half consumed, its jaws working rapidly.

An insert showed the monkey in trouble, as it had accidentally squeezed and crushed the cup-like half containing the balance of the egg and was gathering every part of this by licking its hands and dexterously darting its tongue to both sides of its mouth. The grasshopper won the contest, with the monkey a close second.

When I projected this sequence of scenes, Andy shouted with glee.

"You've got the idea, doc!" he exclaimed. "That's the best yet."

Andy was right. There was no need of peddling

these reels. They were disposed of to excellent advantage, and to my surprise there were but slight projecting room suggestions. When the reels were put on the theatrical circuits I dropped into the New York Strand one evening to see how an audience liked my pictures. The waves of laughter that swept through the big theatre were mighty heartening, even though I was a scientist producing something that wasn't scientific.

That was the way my film laboratory was put on its feet. I was able to make extensive series of technical studies showing life histories of insects, development of frogs and salamanders, and studies of strange types of reptiles. Then I prepared another batch of comedies, and with the funds continued the scientific recording of the habits of mammals, including monkeys, and the engineering habits of the beaver.

Sets of reels at last shaped up for lecturing. I had long anticipated presenting lectures with thoroughly worthwhile illustrations. There was much praise about the character of the films and it was gratifying to hear the words "wonderful" and "remarkable" quite generally expressed. In several large auditoriums, however, where there were mixed audiences, I sensed a certain heaviness during the passing of an hour and a half in which I showed six reels of insects, or talked solely about reptiles. The trouble seemed to be a too academic handling. I felt that if descriptions could be simplified, the pictures would be more interesting to a general audience that included many young people. But I was wrong. Andy again solved the problem, but when he presented a radical suggestion, I wouldn't listen to him. It seemed too outrageous.

Andy had accompanied me to one of the high school lectures. When it was over and we were riding back, he spoke out.

"Same old trouble, doc. Your arrangement is all wrong!"

"What do you mean? Do you think I'm going to *lecture* with animal comedies?"

"No, no," assured Andy. "But you can't hold 'em over an hour on bugs and spiders, nor fill in that much time on frogs and snakes. Some of 'em will go to sleep on you. What they want is variety!"

"What do you mean by variety? There were the habits of fully twenty species of insects shown on those reels tonight. Don't you think they were interested in the way those caterpillars spun their cocoons—and the emergence of the moths? Those were remarkable films."

"Great stuff all right, but too much of it. Now if you'd had a reel of monkeys, one on beavers, another on snakes, one of the bug reels and so on, they'd have given you a big hand."

"You mean an animal variety show?" I demanded, angrily gripping the wheel of the car.

"Now listen, doc, don't get sore. What I mean is this. The monks act smart, but are they? The beavers put it all over the monks. There you are, two reels. You show the beavers are workers and the monks plain loafers. Then you show that the caterpillars put it over on the beavers by spinning in silk, and that's three reels. Then you come along to your poison snakes, which think they have everything stepping, and they get knocked out in a few rounds by your smart mongoose and hedgehog and secretary bird. That's four reels and you can easily work in another. There you are, and as you know the whole works about these things, with a good line of talk, it'll go over strong."

I was sitting stiffly erect, but not in anger. Andy had suggested an idea filled with possibilities not jarring to me as a scientist.

"Comparisons of habits," I was muttering. "Comparisons of the habits of the higher types with the lowly. The building of homes without hands. Structural work of mammals as compared to the details of insect shelters, traps and snares. Parallelism in habits. *Parallelism—*"

"What's that?" asked Andy, as my thoughts had become audible.

"Wait and see," I retorted. "You'll be satisfied."

At a lecture in New Jersey, with fifteen hundred in the audience, I tried out the new idea. Andy watched the show from the projection booth, where, I suspected, he bossed the operator throughout. The lecture was a great success. I received the "big hand," as predicted. We were both elated.

Despite my early antagonism to advice from the movie fraternity, I have come to believe that some of their suggestions are of much value, even to a scientist. However, there are times when I revert to the ultra-academic, in scientific institutions, where the audience will stand for it, and thus ease my conscience from lingering qualms about the Jungle Circus.

## CHAPTER VIII

# MAMBA

In my laboratory film work, making scientific movies, the most thrilling experience I have had was with a green mamba, that whip-like and most dreaded of the serpents of Africa, headquarters for the world's worst assortment of reptilian deviltry. There were times, during that experience when I thought I was going to "get it," but as I had stuck my foot in it, so to speak, I had to risk the consequences.

Africa is the home of big vipers and a dozen kinds of cobra. Among the latter there have occurred examples of what scientists call specialization. Some of the cobras have developed the habit of spraying poison from their fangs toward the eyes of an intruder—and their kind of venom is immediately absorbed through mucous membranes. Even the moist portions of the lips are susceptible to it. In Lake Tanganyika there is an aquatic cobra, which grows to a length of at least eight feet, and will spread its hood and strike while swimming. The most specialized, however, is the mamba, which has switched from the habits of its ancestral stock, become arboreal and taken on an excessively slender outline. It grows to be ten feet long, but one of such dimensions is not much thicker in its middle part than a man's thumb. Its head is slender, like that of a harmless serpent, and to add to the deceptiveness of its appearance, its eyes are large, with gilded iris and round pupils. It is the antithesis of the awful looking vipers—the Gaboon, puff adder and nose-horned species—which roam the forest floor of the mamba's realm. Those viper types, adult when four feet long, and as thick in the middle as a fire hose, have heads as broad as a man's hand.

Precisely speaking, there are four recognized species of mambas, much alike except to the eye of the scientist. Colors among individual kinds vary from green to olive and brown, and some specimens are black. Mambas are to be found over the greater portion of tropical and southerly Africa.

The chief point about mambas is their alleged disposition to attack human beings with a bewildering rush from which there is little chance to escape. Shuddery stories tell of the downward rush of a lithe form, as quick as a released cable sliding through branches of trees and bushes, a deliberate bite, then a gliding rush away. The second outstanding point about mambas is the declared deadliness of their poison, although the poison-injecting fangs are small—barely a quarter of an inch long. Such fangs do no more than exude a few small drops of almost colorless poison. Here again is a great difference between the spurts of yellow venom from the inch-long, hypodermic teeth of the vipers—but the effect of a mamba bite is said to be the more deadly.

While I have not explored the home area of the mamba I have had my ideas about it. During years of caring for the mammals and reptiles in the Zoological Park I have several times received mambas. One batch gave the head keeper and me a shock by going into action like released springs, just after we had worked them into the big exhibition cage, and before we could slide the door shut. One had shot over our heads, like an elongated arrow, and seemed to vanish. A moment later we saw him looking over the side of a shelf, close to the ceiling, with open mouth. The tiny fangs, which looked like splinters of glass, were at the very front of his mouth, where a mere nip would have imbedded them. Through slow movements with a long pole having a blunt hook at the end, we induced the reptile to utilize the hook as a stepping-stone to cross the passageway and enter the cage.

This show of activity had not changed the idea I had formed about mambas. I did not regard them as approaching in insolence or aggressiveness the king cobra, another of the few allegedly attacking types of snake. My idea of the mamba's habits was that it could be compared to the American blacksnake; that

AFRICAN GREEN MAMBA

The scene is an enlargement from movie film produced during one of the most thrilling times in the studio.

AFRICAN PUFF ADDER

Its excessively stout body and broad head illustrate the great difference between the viper and cobra types of venomous serpents—the mamba being a tree cobra.

HEAD OF MAMBA

Slender outlines and large eyes impart a particularly innocent picture—for a snake—yet here is one of the world's most deadly serpents.

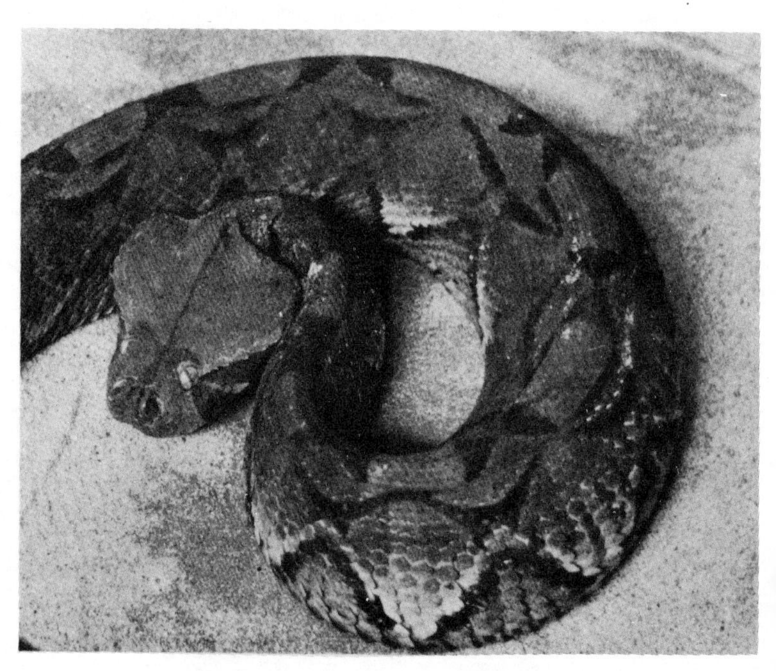

HEAD OF GABOON VIPER

The villainous aspect, huge fangs and large amount of poison combine no greater danger than the nip of the small-fanged mamba.

it did attack, at times; but that such attacks were largely limited to the breeding season.

I have several times been attacked by blacksnakes when up on the sunny mountain ledges in the spring searching for dens where rattlers hibernate. They have glided towards me and struck as high as my knees. One startled me by crawling on my back while I was peering under a flat rock. Another was a nuisance in following me from one spot of investigation to another. I finally got him by the neck and tossed him over the edge of the ledge, fully fifty feet. He landed on the springy tops of sumac bushes and was not harmed. Such occurrences are not common. I have seen dozens of blacksnakes during the breeding season on a spring hunt and every one of them streaked away. The next year, one or two bold and irritable individuals might be met. Then several years might go by without my seeing such a specimen. This, it seemed to me, might be the story with the mamba. There are plenty of mambas in Africa, and exciting tales about animals travel quickly, and radiate broadly, even over continental areas. A few attacks by mambas, where white men were concerned, would give the snakes a sinister reputation. Guided by this, I took one of the mambas to my studio, where branches had been set up, and made several good scenes of it, one showing a waving of the head characteristic of tree snakes. I moved slowly and was very cautious, but the reptile was quieter than the average poisonous snake. There was little difficulty in guiding it back into the fiber carrying case.

I told Head Keeper Toomey about it. "Sure," said Toomey. "They're not so bad as they're painted."

Later I received several large mambas, the finest ever on exhibition in the collection. One was about nine feet long, vivid bluish green, with the narrow spaces of skin between the scales an inky black. This made the relatively very large mamba type of scales look like parrot feathers. I decided that this particular kind, captured in the Angola district, must be recorded for the reptile files. Setting the fiber carrying case in the exhibition cage as a trap, I gently poked the largest specimen until he started moving about, and in he went.

For this fine snake I took considerable pains in the studio, having cut down a young hornbeam tree, which, with its horizontally spread branches, would give the mamba a chance to lie extended and look down at the camera. In its natural surroundings the mamba lives three-quarters of the time in a tree. Behind the tree was a big panel of neutral tint. The lighting was by mercury vapor units suspended above the tree, with the beam of a spotlight from the side for "modeling."

I had been carrying on alone in the night work of the studio for some time, as recording was now intermittent and for the purpose of filling gaps in the photographic series. Through a friend in a big film laboratory

I had landed Andy a good job. However, I preferred to work alone in any case with poisonous snakes. The less movement of human forms the better. Besides, the lights now functioned automatically, and the camera was run by a motor controlled by a flexible cord snapped to my belt.

I unlocked the fiber case and pushed the cover back with a bamboo staff, which carried the usual blunt balancing hook at the end. The mamba reared from the case as straight and to the length of a walking cane. It was looking at the radiance of lights. It rose higher, until fully five feet of its body were almost vertically extended, when it swayed slightly, from side to side. This was my cue to guide it to the tree by placing the blunt hook against its undersurface. It responded by forming a slight angulation of its slim form, and with this resting on the hook, I lifted it toward a lower branch of the tree. Seeing this, it pushed forward and upward, gained the tree, and now with every part of its whip-like body in motion, distributed itself horizontally into an ideal positión for photography.

As I sought to extricate the stick so as to get it out of the scene before starting the camera, the twisting of the hook caused it to slightly touch the serpent's tail. While a moment before that same hook had assisted the reptile to its perch, it had been confused by the lights, probably not realizing that the stick was being manipulated. But stimulation by the lights had been almost instantaneous, and that brushing contact with the tail was a startling illustration of this. All portions of the serpent were again in motion and as I bent down and moved back, I saw that part of the movements formed a big double S in a lateral plane of the forward part of the body. Before I was back far enough to save myself a grunt of concern, the blue-green form shot at me so that I ducked to the floor. The head probably missed me by a yard, but I felt a pulse beating somewhere in my throat. Here was a reptile warranting great respect. The loops were formed again and the scene was so impressive, that right where I was I pressed the contact button on the electric cord, and the camera recorded the pose. By waving a hand slightly I caused the serpent to turn its head, and so look down at me with slightly flattened neck. This was the moment for another scene, to show cobra relationship in forming the trace of a hood. But it also hinted at rising anger. My idea had been to change lenses finally so as to get a close-up of the head. But I changed my mind. The two scenes were enough.

Such transition of temperament among serpents is not unusual. I have noticed it with specimens long captive, when taken out of doors into a genial, spring sun. I have seen reptiles we called tame take a few breaths, and a moment later flatten out and strike. The mercury lights, rich in ultra-violet, had affected the mamba like spring sunshine. So here I was, learning something about bringing a belligerent disposition to the boiling point, coupled with chain-lightning speed. There was now not the remotest chance of "coaxing" this particular serpent back into the fiber case. He had it in for me, and I was earnestly wishing he was back and locked up. I kept thinking of those little fangs up near his nose, like slivers of glass, and even of some of the stories of how a mamba could descend from a tree as if it were spilled out of it.

I sat quietly on the floor and thought it over. How could I get him into the case? Had the rays of the lamps stimulated that brain to the extent that psychological wires were crossed and the radiance indicated Africa, during the mamba breeding season? If so, I was in for an unpleasant time, and would have to step lively. However, there was no uneven forest floor on which to stumble, but the cement surface of the studio, coated with floor enamel, which reflected the lights like glass. No, my elongate adversary would have to swim over such a surface if he followed me up.

My decision was to argue with him—but with a longer pole. There was one at the rear, used in getting hold of pulley weights, which sometimes shot upward, out of reach. I slid backward along the floor until I reached that pole, then picked it up and walked slowly toward the mamba. He was studying my moves, as I could see by his head being lifted higher, but inclining directly at me. As an experiment, I pushed the pole at him, and touched him on the side. He struck, half his length, an energetic straightening of the loops. As the head reached the limit of the stroke, it turned sharply, and the pole was seized in a chewing grasp. The motions were very different from the flashing jaw action of a viper. They were cobra-like.

This demonstration was followed by a move more disturbing, though not directed at me. The mamba reared his head, then pointed his body straight upward and inspected the overhead part of the studio.

Here was a maze of pulleys and high electric conduits. If the animated whip got up there, I would have a fine time chasing him! But that was just where he went, making contact with a pulley by stretching what appeared to be two-thirds of his body upward. I had never seen such dexterity in a snake, and could only account for it by the mamba's excessively slender build.

With a grace that was fascinating, that nine feet of green traversed the length of the studio, bridging cross stretches of conduits, gliding through ring-bolts and stopping at a spot that made me breathe fast.

Not a yard in front of him, was the open end of a conduit, with a flare terminal. From this, several wires extended in a loop to a suspended mercury lamp. There was ample room for the snake to *enter* that conduit, one of a great many; and all those pipe-like, carrying passages for the wires were coupled in a maze. They led into switchboards and some of them opened to the power lines, out of doors.

That situation sent me on the run to a wall cabinet, where I grabbed a ball of noosing cord kept for snake-hunting trips. I fumbled for my knife and cut off a couple of feet of the cord. My eyes kept flashing upward to the snake. With relief I noted that his neck was doubled back again as he watched my motions. This brought it well back from the fateful opening.

It wasn't easy to raise the pole and get the slip-noose, now fastened at the end, over the mamba's head. The pole bumped ropes and wires, but the mamba, luckily, had now decided the pole was inanimate and didn't strike at it. I managed to swing the noose over his head, and pulled, and nine feet of mamba came swishing to the floor.

The meanest part was to get hold of him. I had to keep the end of the pole upward so as to keep the weight of part of his body on the noose, or he would slide out of it. As the noose was fairly short, I grasped the pole a bit nearer to him than the center, maneuvered until I obtained my light pole with the hook, then edged up on the snake to get the hook across his head. After some sparring that mamba was pressed to the floor. I dropped the long pole and grasped him by the neck.

While for a moment I was somewhat tangled with the protesting body, my troubles were over. I was surprised to find how easy the creature was to hold. There was no more difficulty in restraining him than a common, five-foot black racer. Despite this creature's remarkable aptitude to reach upward as straight as a rod a distance of more than half its length, and to strike fully that distance, its neck felt thin, soft and delicate. A blow with a Malacca walking-stick would have broken it. No wonder Nature had arranged that such a fragile type should distill a venom of highly lethal potency.

With the mamba back in the fiber case, and the cover shut, and locked, I gazed at the receptacle. What had happened? The mamba had struck twice at me. There had been an exciting chase to get him back in the case; that was all. I figured that if Toomey had seen all this, he would have reiterated what he had said before about mambas:

"They're not so bad as they're painted."

It is probably owing to sinister stories about the mamba, the Indo-Malayan king cobra, and outstanding poisonous snakes of various countries, that serpent lore ranks head and shoulders over stories and traditions about other groups of wild creatures. The harmless snakes as well are responsible for strange tales and superstitions. Here appears to be the reason for letters relating to snakes outnumbering all others received in the combined animal departments of the Park. I am constantly hearing more of this lore: some I believe; about a part I am sceptical; and the rest seems to be utterly improbable.

J. L. Buck, of Camden, New Jersey, blew into my office on returning from an African trip. He had brought a collection of rhinoceros and Gaboon vipers, and spitting cobras. Visitors were soon flocking about the cages to look at the astonishing and bizarre patterns of the vipers. One of the cobras, a big fellow, was much excited at first, and spent the time rearing two feet high, with widely spread hood. This serpent was lustrous blue-black, and on each side of its hood was a marking like a brilliant, red butterfly. I was keen to know under what conditions Buck had captured the snakes. He is a skillful collector, has brought many rare and valuable specimens to the States, and invariably his specimens arrive looking as fresh and blooming as if they had just been captured. He explained that the greater number of the snakes had been snared.

Our conversation swung to mambas. Buck said that some specimens attacked human beings during the breeding season. I related my experience in the studio, and my visitor told me that during a former trip he had seen the Kaffir medicine men, or witch doctors, handle mambas with impunity, and had learned the secret of how they could do it.

They brought mambas to him for sale, carrying the snakes in baskets with flap-like covers. The medicine man would raise the flap, reach in and pull out several mambas. The snakes were alert, but showed no signs of anger. Buck heard that the witch doctors, in the first instance, had the same troubles as anyone would have encountered in capturing them, which he thought had been done with a snare on a slender pole. The snakes had then been worked into the basket.

After several mambas had been captured, the medicine man prepared to become so thoroughly acquainted with them that they could be handled with impunity. The procedure seemed humorously simple. It was based upon the effect of an old shirt!

During the middle of the day, when zebras and antelopes pant and men perspire, the medicine man would exert himself enough to reek with perspiration. During these endeavors he wore an old cotton shirt, and this he rubbed vigorously against his chest and sides until it was saturated. Stripping off the shirt he went to the basket of mambas, which had been placed under a bush to protect it from the sun. Operating the flap cover, he pushed the shirt inside. That was all. It was a question of introduction through scent! The next day he could reach in and pluck mambas from the basket without fear of their biting him.

Buck appeared to believe the story, and I was inclined to do so, although expressing myself as sceptical. However, we agreed upon the strange control of recently captured mambas by the witch doctors, when those same snakes, upon arriving in the States, were found to be in possession of their fangs, and as irritable and vigorous as any. We discussed the seeming immunity from harm of some Arabs and Hindus who could handle cobras, which appeared to be calmed in the men's presence. I had not found a solution for such manifestations, nor been able to quickly "tame" poisonous reptiles, myself.

Not long afterwards, the story was substantiated, and right in the home area. Head Keeper Toomey met me as I came into my office in the morning, and asked me to come to the front of the python cages.

In the central cage was an Indian black-tailed python, about fifteen feet long.

"Where did *he* come from?" I asked Toomey. "Somebody make us a present?"

Toomey said he had been in a predicament the night before. A taxi had rolled in just after I left. There was a trunk beside the driver. Alighting from the cab was a circus snake charmer, who had often written and telephoned me for advice about her collection of pythons. They were her stock in trade and received scrupulous care. She told Toomey that while waiting for the opening of the circus, she had secured a week's engagement out of town and couldn't take the big snake with her. He was nearly ready to shed his skin and she couldn't leave him shut up in his trunk in the apartment, as he might go to sleep and the skin would harden on him, giving her the tedious job of soaking him in the bathtub, and then peeling him inch by inch. So she implored our good offices in keeping "Rajah" in one of our cages until she came back.

Toomey was in a quandary, as the Park has rules against accepting specimens for deposit. He refused to take the snake and the lady shed tears. "Rajah" would be no trouble, she insisted. He would require no food and would not take up a minute of our time.

As Toomey is kind-hearted and felt that to turn away the weeping lady and her trunk with no place to go would be on his mind, he helped the taxi driver in with the trunk.

The snake enchantress said she preferred to take "Rajah" from the trunk herself. Toomey didn't object when he saw the size of the brute, with whose possible vagaries he was not acquainted. It was a black-tailed python. That particular kind has a pretty uniformly mean disposition. He was surprised, in fact, to see the lady in possession of a python of this sort. The good-natured rock-python is the kind generally used by "charmers" in the shows.

The python's owner, however, reached into the trunk and placed an upturned palm under "Rajah's" chin. The python had a head as big as that of a terrier dog. Three inches of a purplish tongue waved nervously. Toomey didn't like the looks of the brute and suggested that he get a horse blanket, throw it over the snake's head and grab the creature by the neck, while the lady and the taxi driver handled the balance of the reptile. Even at that, he figured, they would have a tussle. Fifteen feet of heavy-bodied python was a fair amount of snake, if it objected.

But the lady would have none of this. She was able-bodied, really statuesque, and after stroking the python under the chin, proceeded to hoist its anterior part from the trunk. It would have been a staggering load for a man, but she draped the forward third of the snake over her shoulders, took a step toward the cage door, then turned her body so that the python's head inclined toward the floor of the cage. Her left hand guided the heavy length now issuing from the trunk. The creature went into forward motion, and into the cage.

As Toomey closed the trunk, he noticed something inside, a garment of some kind. The lady followed his glance.

"I always leave something of mine in with him," she remarked, "to remember me by."

When I saw the python the next morning and these details were discussed, I remarked to Toomey that this was a rare example of a black-tailed python proving so docile.

"Docile!" exclaimed Toomey. "He's a wicked brute. He made a shot half across the cage at me this morning."

I ruminated about this and the garment, then went to the trunk to look at it. It was of the type, which I believe, is called a slip.

Unfortunately, the lady called and removed her python while I was away. I meant to question her about that garment. Since then she has been traveling in the South, and I haven't had a chance to pursue further investigations about this garment theory of snake charming. However, I am going to try it some day.

I will confide to the reader that a lot of people think I have an uncanny influence over snakes. I know how to handle the poisonous kinds to the extent of having kept out of trouble thus far. In years of handling snakes, probably hundreds of them, I have never been bitten. But I have no uncanny influence, and the realization of this has, at times, given me much apprehension, as I will explain.

Occasionally I am called on to do considerable snake handling outside the laboratory. Some lecture audiences demand that live snakes be brought along. Such lectures are usually before medical societies, where I have been asked to explain methods of extracting poison to be used for investigations in the treatment of cancer and abnormal bleeding. In past writings I have done no more than touch upon this work, as we were but at the portals of a new phase of science. In the next chapter, however, I shall for the first time go into details of what I feel is the most important work with which I have had contact and to which I could extend co-operation. At any rate, these matters have brought about a keen interest in living serpents among the medical associations.

The snakes are carried to such gatherings in a large satchel, each reptile in a separate cloth bag, properly tagged. I usually take a fair variety of harmless snakes, and for poisonous kinds a rattlesnake, a copperhead and a southern water moccasin, the latter known in its home country as a "cotton-mouth." This latter kind is particularly satisfactory in demonstrations of extracting venom. If carefully handled, on the speaker's table, it seldom becomes hysterically angry; yet it is indolently insolent enough to stand by you and fight, and yields a relatively large amount of poison to show up in the glass.

The method of procedure when arriving at the discussion of the poisonous snakes is simple, and I have never had anything happen to cause me to change it. In the satchel is a deep canvas bag, with three pieces of cord tied to the mouth. I use these in tying the bag open against the backs of two chairs, producing a triangular opening at the top. When I come to the poisonous snakes, I first slide the rattler from its traveling bag to the table, where it sounds its rattle, and the audience is impressed. During that demonstration

I show how a poisonous serpent is noosed. I have had rattlers slide off the table and had to noose them on the floor. This has happened when the front row of chairs at a medical dinner was too close to satisfy me. It is often difficult to regulate such conditions. The audience has placed too much reliance on my control of the snakes. Fortunately, I am not affected by nervousness before audiences. If I were I should consider it gravely reprehensible to use poisonous snakes as illustrations—and no venomous serpent that I have ever carried was in anything but full possession of its fangs. I would feel like a traitor to the snakes if I tampered with their defenses as provided by Nature. In times like that I have forgotten the audience, except for the proximity of the feet in the first row of chairs. For a few moments we live in a small domain, the snake and I, and the former must be restrained from getting out of it.

I remember an affair where I operated from the center of the guests' table after a big dinner. Dr. Nicholas Murray Butler sat on one side of me and Ambassador James W. Gerard on the other. I should have preferred more room, but Dr. Butler, presiding a few weeks before, as president of the Union League, had watched me go through a similar lecture, and sat quietly assured. Mr. Gerard took his cue from Dr. Butler.

As the respective snakes are disposed of after such

demonstrations, they are dropped into the deep bag fastened against the two chairs. Here they coil together as the viper types usually do, in a mutually watchful fraternity. They seldom try to ascend the bag. However, I watch that bag intently, until the last one is in, and it is time to tie it up.

The satchel carrying such a collection is usually transported in my car, as most of the lectures are in the area surrounding New York City. Recently there was a lecture before a New England society, and the transportation of the satchel involved difficulties.

Not long ago I had transported a king cobra in a sole leather carrying case, the serpent being in a canvas bag inside, to the National Zoological Park. It was a rush job, as the serpent was needed for the opening of a new reptile house. The satchel traveled with me in a sleeping car. The president of the sleeping car corporation heard of this and sent me a letter of objection, forbidding me to travel again in a car of his company if I repeated the performance.

The New England society necessitated transporting a satchel of snakes, and as a sleeping car was involved, I sought to respect the points raised in the letter. I blew into Grand Central Terminal a half hour before train time and took the satchel to the baggage counter, asked for the excess rate on the receptacle, and told them what it contained.

To my surprise, they refused to accept it, and after

stepping up through subofficials to the czar of baggage transportation was met with a final refusal.

With an eye on my watch, I transferred operations to the night express department, having ascertained that an express car was going through to Boston that night. To my astonishment, the express people also shied away from the satchel and refused it. I countered by telling them that if those snakes didn't arrive in Boston in time for my lecture, I would take action against their refusal as carriers. Their objections were to the satchel. They insisted that snakes should be shipped in a *cage*.

With ten minutes to get back to my train, and in the turmoil of a pre-holiday rush, there was still much fussing. They said they had no pasters sufficiently large to mark the bag properly, but somebody came with a large and ornate paster that was duly plastered on, and I rushed to get aboard.

Arriving at Boston and glancing at the paster to solve its ornate decorations, I found it bordered with an undulated margin of holly; and in bold lettering, struck obliquely, was the advice: "Not to be opened until Christmas."

CHAPTER IX

# PUTTING IT OVER

DURING the twenty years that I have been lecturing I must have talked to about half a million people. As a matter of fact I have kept a note book to remind me of the places visited and the size of the audiences. I have never enjoyed using lantern slides, because the "still" image of an animal staring out from the screen has never seemed to me convincing; but they were necessary in the early days to supplement my living exhibits. At first I used to take a selection of reptiles. Later, when talking about warm-blooded animals, I carried small specimens in light cages. Head Keeper Charlie Snyder and I would set out at night in a big seven-passenger touring car to give lectures in school auditoriums. By taking out the seats in the rear we could carry a big trunk containing the cages. Another case held a combination spot-light and stereopticon. We also took a folding table. I would talk about the structural forms of small animals, their habits and their natural surroundings, and illustrate the first part of the talk by lantern slides which Charlie "shot" from the stereopticon. We then set up the long, folding table, and Charlie slipped the front from the stereopticon and converted it into a spot-light. One by one I liberated my specimens from the small cages and kept up a running commentary, while Charlie swung the spot-light on them supplementing the lights of the auditorium. The table was equipped with several devices on which the animals could climb and show themselves off to the best advantage. I had a fine collection of small animals, each tamed to remain where it belonged, so I felt fairly sure there wouldn't be any commotion. However, I was never entirely free from a jumpy feeling that something might happen—and once it *did*.

My collection consisted of several groups of related animals. There was a woodchuck or groundhog, a tropical squirrel, a gopher rat, a pack rat, a porcupine and a kangaroo rat or jerboa from Africa. These were the rodents. There was a tame skunk, which had been rendered harmless by the removal of its scent glands, a weasel, a coati, a kinkajou, a Madagascan tenrec, two species of small monkeys, a Bornean loris and an opossum. What astonished most persons was the fact that I could jam this menagerie into a trunk. The animals traveled comfortably, however, and most of them lived for years after their evenings of public appearances. I rather imagine that the presence of the skunk gave the audience its biggest thrill, but my greatest difficulties came from the porcupine. It was as tame as an animal could possibly be, but its spines had the tendency of showing up, or literally turning up, everywhere. They were to be found in utterly unexpected places and how on earth they got there was a mystery. Once I writhed through part of a lecture with one of them sticking me in the side. Several times we rose in anguish from the cushions of the car. Charlie and I often discussed this drawback about the porcupine, but decided we could not dispense with the attraction. It was one of my smallest specimens, however, that gave us the greatest shock.

The incident occurred at a high school in Brooklyn. I had progressed through my descriptions of all the rodents but one. The woodchuck or groundhog had been explained, and how his appearance and casting a shadow on Candlemas Day was just incidental to the weather being mild enough to lure him out of his burrow at that time. During this explanation the stocky little creature sat up and looked cute, blinking at the light. The palm squirrel had climbed to the top of a notched stick and I had gone into details about there being over fifty distinct kinds of squirrels in the United States alone, several hundred species known throughout the world, and how some of the squirrels in India were as big as a cat. The Texas gopher rat had shown how it stuffed its cheek pouches with food, and the pack rat had demonstrated its craze for collecting a jumble of bright objects for its nest, by carrying my watch fob across the table and taking it into its cage. Slid from its cage the porcupine had erected his quills as I rubbed him the wrong way with a stick. The remaining rodent was the Egyptian jerboa.

This curious little rat is like a miniature kangaroo and runs erect. Its color is pale tan. It has flattened pads of hair on the hind feet so that it can jump and run over the soft surface of the desert. It can move at such a speed that it appears to skim over the sand. On the table it moved back and forth so quickly that its spindle legs could scarcely be seen, but I wanted to show the audience its jumping powers. To do this I propped a partition of cardboard a foot high across the middle of the table, and when the jerboa came to this it soared over the hurdle as lightly as a feather. I was about to remove the pasteboard barrier, when a man in a front seat gave an explosive cough.

This time instead of hurdling the obstruction the jerboa shot a yard into the air. Either the cough did it, or the animal stepped on a porcupine quill. At any rate the leap described an outward curve and I saw the jerboa go sailing to the floor below, where it landed in the center aisle and looked like a yellow streak making for the rear of the auditorium.

It doesn't take anything very unusual to produce turmoil in a crowded audience, and while the little animal was of an innocent kind, it was a *rat*, and the female

AFRICAN JERBOA

This stilt-legged rat of the desert can run at great speed and when stopping rests on the feather-like tip of the tail. Its feet have pads of hair to prevent sinking in the sand.

BRANICK PACAS

They inhabit the Amazon Valley, live in burrows, are largely nocturnal and seldom obtained. Few have been exhibited.

AFRICAN SPIKE–TAILED LIZARD

Compared with the creature figured beneath there is an interesting contrast in tails. There is great diversity in shape and length of tails among lizards.

AUSTRALIAN STUMP–TAILED LIZARD

Looking like an exaggerated pine cone it is difficult to tell whether such reptiles are coming or going.

percentage of the audience was large. The racing of the creature along the table, my insistence that it could go even faster, and that final show of its leaping straight upward, had made the atmosphere tense. There was a general murmur which increased in volume, and some louder sounds of agitation.

I can't remember what I was saying. Possibly it was to the effect that a canary was savage compared to the gentle jerboa. What I do remember was that my gaze concentrated on that yellow streak. The volume of sound was building to a crescendo and almost drowning my voice when I saw Charlie step from behind the spot-light. As the speeding animal reached him, he bent and made a sweeping catch. His hand closed around the jerboa and he held it up in the beam of light. There was a spatter of applause partly muffled by a concerted sigh of relief. Charlie came to the platform and we made quite a gesture of putting the jerboa back into its cage and snapping on the catch. I decided to say nothing about the skunk, and to keep it out of sight, harmless though it was.

Those lectures meant lots of work and worry in transporting the animals. I had to keep them warm on winter nights, and managed it by the use of hot water bags. But I was waiting for a time when I could afford to make motion pictures of large and small animals. The funds from those talks, which fortunately proved popular, built the first section of the laboratory.

Now, in a metal case of reels selected from about a quarter of a million feet of film in the vault, I can take along anything from elephants to cobras, and be assured they will behave.

In my early twenties the prospect of facing an audience made me pretty nervous. By the time I was thirty I was addressing a thousand or more people with no feeling of stage fright. This sense of confidence as well as a feeling of pleasure in lecturing, comes, I think, from care in arranging illustrative material and a sympathetic knowledge of my subjects. Experience has taught me to study the length of each scene in each reel of film used with a lecture, and to time my narrative so that the flashing on of a new scene synchronizes with what is being said. The mere changing in brightness of the light reflected over my shoulder from the screen gives me the cue, and there is no need for me to turn my back upon the audience to look. To use Andy's phrase, it is a case of "putting it over," and a duty of the speaker if he has any respect for his listeners.

A good or bad motion picture operator can make or break a lecture, and a speaker must be prepared for emergencies. I have visited small high schools where there was a projection room as well-equipped as in a theatre, and where a licensed projectionist ran off the reels in sparkling fashion. I have also found poor machines and incompetent operators. In such cases the reels are apt to be torn, scratched or broken, and run either too fast or too slowly. Lecturers on lengthy tours, like Martin Johnson, take along highly efficient operators and up-to-date projection equipment. I have employed Johnson's operator at several places, and his work was flawless.

The gaps between illustrations, when operators or machines go wrong, are maddening to a speaker, but through years of experience I am prepared to smooth them out. The talk must go on. I carry in mind a long list of stock experiences to fill in the time, but there have been occasions when I was so tired after fighting against the bungling of a poor operator that I felt like crawling away and taking a nap—after spanking the members of the arranging committee.

The most extraordinary demonstration of what can happen when the projection goes wrong occurred in a western city. There was a semi-portable machine set up in the aisle. I didn't like its looks, but the pictures went along for a couple of reels. Then I heard the mechanism rip out the perforations on the edges of the film. The machine was stopped and I could see that the operator was doing something. After a few minutes he had it going again, but the picture danced and the tearing of the film began once more. I could feel my forehead breaking out in a sweat at this punishment of my reels. Then I heard the clank of tools. Here was a new angle. The operator was going to overhaul the machine, while I struggled along with the lecture.

His efforts were brief. To my astonishment I saw by the glow from the exit lights that the husky operator had picked up the machine and was staggering out with it. It was fastened to a table, and the whole thing had been lifted by his getting under it and taking it on his back. I recollect that as he moved off bearing the table surmounted by the mass of the machine and with the metal legs protruding, he gave the effect of a small camel leaving the place. There was nothing to do but keep on talking, after telling the audience that we might hope for the operator's return. I had talked for half an hour when the silhouette of a different-looking apparatus came wobbling down the aisle. Within ten minutes a picture was again shot on the screen, and the lecture thence proceeded to an end, three quarters of an hour beyond its allotted time.

Usually under such circumstances there are keen members in the audience who come backstage and congratulate the speaker on his filling in the break. But unfortunately many members of an audience blame poor performance in projection on the lecturer's reels. They never see such things happen at a theatre and seem to think that professional picture film would not behave this way. I once had a tilt with a lecture committee, which complained that my pictures were dim and that they could not see everything that was going on.

It happened that the auditorium was big and the lecture delivered in the afternoon. There was a row of huge windows on one side. They were shielded by velvet drapes, which when pulled over the panels left margins for daylight to come through.

When I looked up at the balcony a hundred and fifty feet away, where a temporary machine had been set up, I foresaw serious trouble if a bulb light was being used.

On inspecting the machine I was chagrined to find that not only was there a bulb light, but one of only five hundred watts. A thousand watt bulb would have barely given fair illumination under the conditions. The operator, an instructor in the high school, said they had decided the machine would do. The committee indicated that I was butting in; that it was my part to furnish reels and theirs to provide the machine—and this was a moving picture machine, and what more did I want?

The first dim glow on the screen projected from that distant balcony, with daylight seeping in from the windows, convinced me that the audience must have sharp eyes to gain the significance of the pictures, and I told them so. Then, the best I could do was to throw all my energy into the talk to neutralize, to some extent, the depressing screen effect. I finished that lecture in a reek of perspiration—for I had *worked*. In the face of this I was met by the committee, who objected to "poor" pictures.

There were five of those criticising gentlemen. I was so enraged at their lack of understanding about the principles of light, that I insisted we go to one of the local theatres and see if it could be arranged to have one of the reels run under proper conditions. They demurred, but I told them that in fairness to me they couldn't sidestep.

In a cozy little theatre near by, we were ushered into the manager's office. He said he would slip in a reel, as it was between times, but asked if I would explain it to the audience. We took the reel to the operating room and a lean young man, keeping an eye on one of the motor-driven projectors, listened to the story.

"I happen to know what I'm talking about," I told him, as I showed my card as a licensed operator, issued in New York. He warmed up.

"How long was the throw?" he asked.

"Fully a hundred and fifty feet."

"And they were using a five hundred watt lamp?"

"Yes."

"Twenty-five hundred, from an arc, would barely do it!"

The committee went downstairs and sat in orchestra seats. I waited in the wings until the reel flashed on, then stepped out and talked through it. No dramatic reel could have shown more brilliance and detail than those scenes.

The event closed happily. The committee took me to a club for dinner and admitted they would henceforth seek professional advice before installing a motion picture machine.

There are many kindly people who invite the lecturer to dinner and others who invite him to stay over the night. As a rule I try to sidestep dinners unless with someone directly concerned in handling the lecture. I never stay in a town overnight if there is a chance of moving on toward home in a sleeper, as it saves valuable time. While I prefer to go to a hotel room and rest before a lecture, this nowadays is difficult. If there are no dinner arrangements the local papers want stories, and their representatives visit the hotel room. Another thing the modern lecturer encounters is the radio talk. It is becoming a regular thing in smaller cities for the lecturer to be requested to give a short broadcast during the period before dinner time, on the evening of his lecture. This is supposed to increase interest in the town, and if the lecture is in connection with an educational institution the speaker may be asked to say something about the organization's work. In the very small towns, and it sometimes seems they have the finest auditoriums, everybody turns out, and if the lecturer is not leaving before a late train, he is taken in hand after the show and driven to the home of one of the town's influential citizens. Here he meets a gathering assembled in his honor, which usually includes the mayor, president of the board of education and other important people. In such instances, the lecturer, of course, is made the lion of the affair, and answers so many questions that he virtually gives a second lecture. However, I enjoy these parties, with cordial and sincerely interested people. If the host is a staid church member, coffee and cake may be served, or if not sabbatically inclined, the refreshments may be spirituous and varied and the lecturer fortunate if he doesn't miss his train, to which he is escorted in a thrilling dash against minutes.

I remember a formal dinner where I was the guest speaker and had brought along several reels of movies. During the banquet the toastmaster, next to whom I was seated, asked me so specifically about the scenes in my reels that I thought he was afraid of the effect of snakes upon the diners, an association of elderly gentlemen. I assured him that while my reputation was associated with snakes, none appeared among my movies, and reminded him that in correspondence regarding this banquet it had been specified that no snakes should appear. Then I asked him if his organization was particularly susceptible to anything of the kind. He told me that at the previous gathering, one of the members, a physician, had volunteered to give a movie talk and brought a set of reels showing surgery on the battle field. The scenes worked up to a point where two of the diners simultaneously collapsed, and there had been requests for a more cheerful lecture at the present gathering.

Some of my most attentive audiences have been in prisons. I gave my first lecture at Sing Sing over twenty years ago, and last year, at the request of the committee representing the inmates, gave two lectures. The question period following a prison lecture always impresses me. The queries indicate careful reading and serious interest in scientific subjects.

One other kind of audience should be mentioned—and I come in frequent contact with it. This is the strictly scientific gathering, composed of medical men or the members of one of the scientific societies. Here the presentation is in the form of a "paper," although I illustrate such talks with motion pictures. A talk may necessitate the complete rearrangement of old reels, or preparation of new ones, going deeply into details. At the time of this writing, I have been listed on such a program, several months ahead. Work on the illustrative material for this "paper" has intermittently been under way a number of weeks. The illustrations alone will have cost several hundred dollars, and the "paper" lasts barely an hour, but a series of original observations are to be brought out and the scientist feels such effort and cost well worth while as going on record.

Possibly of all places where lecturers appear the most sophisticated is the National Geographic Society at Washington. I have spoken before the Society a number of times. A lecturer regards an appearance there as a season's event, for if he has been on that classic course he is said to be "made." The great Washington Auditorium seats about five thousand, and its great expanse looks as big as an armory. A bank of microphones is attached to the speaker's table. These feed into loud-speakers spaced throughout the structure, as nothing but a stentorian voice could possibly penetrate to the rear seats.

Arriving in Washington during the day, the speaker understands that he is to visit the auditorium for advice. His voice pitch is discussed and an electrician who will manage the control board gives him a test. His motion picture reels are estimated as to length—for the lecture must not exceed an hour and a quarter. At the same time, it must extend to this length. The expert operators control the speed of their machines accordingly. The speaker is not introduced. He is accompanied to the wings of the great stage by Gilbert Grosvenor, Director of the Society, or Franklin Fisher, Supervisor of Lectures. The lecture begins on the dot of eight-fifteen, and at that moment, the representative of the Society grasps the speaker by the arm.

"Good luck," he says.

A moment later and one looks out over a sea of

faces, one of the most critical audiences in America, and the largest a lecturer will face in this country. In tones no louder than one would use in a living room the discourse is started. The voice comes welling back from the loud speakers, the latest devices in amplification. There is no blur or distortion. It is rather astonishing and a bit confusing.

I have said that stage fright has been worked out of my system, but the nearest reversion to it I ever have is before that great audience in Washington.

## CHAPTER X

# BENEFICENT COBRAS

Every year about three million visitors come to the Zoological Park. It is my hope that some day they will realize that the reptiles they see are yielding up their deadliest poisons not to kill, but to cure, mankind. I have watched many people look with awe upon the snakes, and I have heard plenty of hostile remarks about them. But there may come a time when a cobra or a moccasin will be thought of as a benefactor.

About five years ago Dr. Adolph Monae-Lesser came to my office and asked if I would give him some poison of the Indian cobra. I have had various requests for samples of poison, and very naturally I am extremely cautious in granting them. This gentleman, however, was well known as a practising surgeon, and one of the founders of the Reconstruction Hospital. He explained his request by telling me of an observation he had made while serving with the Red Cross in the Spanish-American War. A leper had been bitten by a tarantula, and shortly thereafter he had showed a marked improvement in his nervous condition and a lessening of his pain. Dr. Monae-Lesser had stored that curious fact in his memory.

The doctor had already been experimenting with the viperine snakes, but now he wanted poisons that acted particularly on the nerves. There are two types of snake poison: one that has elements we call neurotoxins that strike at the nerves, and the hemorrhagin type which attacks the blood. The neurotoxins are particularly powerful in venoms of serpents of the cobra type.

The doctor told me his research related to the treatment of malignant growths. As I understood him, his experiments were guided by the theory that certain malignant growths were associated with abnormal conditions of the nerve centers. These nerve centers induced activity in certain cells, which congregated and built up abnormal or malignant tissue.

Now, we already knew that in its undiluted form, cobra venom had a terrific effect upon the nerve centers. In fact, when a man is bitten by a cobra, death is largely produced by paralysis of the muscles that control the breathing mechanism.

My understanding of Dr. Monae-Lesser's theory was that if a person suffering from a malignant growth were given a diluted solution of venom there would be a comparatively mild shock in the nerve centers and an activation of the affected centers that would be followed by a correction of the abnormality—a kind of rejuvenation.

As we discussed the subject, I was startled to learn

of the amounts of poison the doctor required. He had several phases of treatment in mind, and in preparing his solutions it would be necessary to make a series of test injections upon animals. I had long been accustomed to extracting venom from serpents of the viper type. With cobras, however, it is a different story. I had occasionally handled them to assist them in shedding old skins, after they had arrived from India after weeks of traveling in tight boxes. Then there had been an occasional specimen with a sore mouth that had to be treated. They are difficult to handle, and muscular. The polished head slips from beneath the staff, and instantly the snake rears and strikes from the table. Even after the head is successfully restrained and the cobra grasped by the neck, there is a feeling as if one is holding a polished cable that could turn within one's grasp, enabling the flexible jaws to be bent sideways and bringing the fangs into a position to grip. If a cobra can imbed its fangs, it *chews*, and in this way it differs from most venomous serpents.

However, I didn't bother the doctor with such details, but was naturally led to discuss conservation of poison. I told him it would be impracticable to have cobras pierce the parchment tied over a glass as had been my practice with the vipers, whose fangs are long, like hypodermic needles. With cobras the poison-conducting teeth are short and stout, and much venom would adhere to the parchment and be wasted. So a small slab of glass was designed on which a cobra could chew. The resulting circle of almost colorless, viscid fluid could be dried and scraped off. Later, the flakes could again be rendered fluid by dissolving them in distilled water.

So the doctor was provided with poison and spent months making test injections upon animals. I was impressed with his cautious approach to the testing of his theories on a human being.

With his knowledge of how cobra venom acted upon the nerves of animals, he at last resolved to use a highly diluted solution of the poison on a patient suffering from facial neuralgia of the left side, resulting from a lymphosarcoma (a type of cancer) of the left tonsil. At the time he had no thought of markedly alleviating the cancer, as it was well advanced. His object was to relieve the pain produced by clearly indicated nerve pressure.

Nevertheless, he was agreeably surprised, for not only was his patient relieved of pain, but the tumor began to diminish. Before the treatment the patient could eat no solids and could not lie down without spasms of the throat; now he could swallow without pain and sleep in a bed instead of on a chair.

That experiment was carried out in Paris where the doctor had taken a good supply of the poison, and it appeared to him and the French doctor working with him, Dr. Charles Taguet, that venom might be used for the treatment of tumors in general.

A year after the first experiment on a human being in October, 1930, they broadened their researches through the cooperation of Professor M. A. Gosset at the Salpêtrière Hospital in Paris, who put many patients into their care. The results of treatments of one hundred and fifteen patients injected with venom during a period of two years and a half were announced to the French Academy of Medicine in a paper which Professor Gosset read on March 14, 1933, Dr. Monae-Lesser being prevented by illness from presenting the facts himself.

For their clinical work the doctors made no selection of cases. They took serious cases, cases recently operated upon, cases operated upon long ago with or without relapse, inoperable cases, patients treated with X-rays or radium without success, visible cancer or abdominal or axic cancer.

"The majority of these patients have died on account of too far advanced cachexia [general ill health and malnutrition due to a chronic constitutional affection], but we have almost always given them relief," they reported. "With others we were able to obtain a stabilization of several months, and again with others we were even able to obtain regressions and even a few complete cicatrizations."

The formal paper cited numerous clinical cases of

various types of cancer that were, to say the least, greatly ameliorated by the injection of minute doses of cobra venom.

"It is generally between the fourth and fifth injection, made every third or fifth day, that the pain begins to diminish, above all losing its acuteness," the paper reported. The conclusion was: "We consider that the treatment by the venom of the cobra will give ease to all pains and particularly to the pains of benign or malignant tumors, and that the latter are benefited by it."

Some of the patients suffering from carcinoma and operated upon by Dr. Monae-Lesser more than five years ago, who underwent subsequent treatment, have thus far had no return of the disease. The time, however, is not considered long enough to warrant assertions or promises of a possible cure. But the treatment is looked upon as of sufficient importance to be tried in suitable cases, particularly those that are not in the last stages, although it has been demonstrated to give relief in most cases.

While Dr. Monae-Lesser's research on cancer treatment has been in progress, another physician has been working with snake poison along entirely different lines. I have watched the development of those experiments too, from my own standpoint of supplier of venom.

This other phase of research is being carried on by Dr. Samuel M. Peck. He is experimenting with the type of venom that contains hemorrhagins, those elements that destroy the red cells in the blood, break down the walls of the blood vessels and generally destroy the tissues. That is quite a different type from the neurotoxic cobra poison. Water moccasin venom seems to be about the best.

Through my possession of quantities of moccasin venom, I have been aware of Dr. Peck's experiments since he began them in 1930. I think it all started when he tried to produce an experimental purpura in rabbits —purpura is a hemorrhage of the skin. He was using poisons obtained from fungi and they didn't work, so it quite naturally occurred to him that a hemorrhagic snake poison of the type that breaks down the blood vessels might be used to make the rabbit's skin more sensitive to the bleeding reaction he was trying to produce.

The fact is that in research one doesn't always get the result one expects. Dr. Peck injected small amounts of moccasin venom, one tenth of a cubic centimeter of a one part in a thousand solution in saline, and then discovered that if he waited ten to fourteen days before trying to produce bleeding, instead of making the rabbit more sensitive to hemorrhagic reactions, not only were the skin hemorrhages lessened, but so was the general bleeding tendency.

That looked like a promising alley to follow, and he followed it. Along the way he discovered a number of interesting things. Whatever it is in venom that produces a resistance to hemorrhage isn't identical with the neurotoxins or the hemorrhagins. These factors—to call them by a general term—are not found in all hemorrhage-producing venoms, because rattlesnake poison, for example, is very poor in them; even individual moccasin venoms vary in their content of these factors and some are stronger than others.

Those were some of the things that research brought out, and after Dr. Peck "titrated" his factors, that is, analyzed them in such a way that he learned the potency of his doses, he started some clinical work on human beings.

Up to the present I understand he has treated about one hundred and fifty cases under very careful control. Roughly speaking, his cases have fallen into two groups.

The first includes patients who are bleeding without any pathological change showing up in the blood to account for it, or any apparent organic reason. Such cases are well known. Doctors call it "functional bleeding," and the two commonest examples are recurrent nose bleeding and very excessive menstrual bleeding.

In many of those cases the loss of blood was so serious that it would eventually have resulted in death if not checked. Dr. Peck found snake venom was a very useful thing. A great many of his patients were entirely relieved after a series of injections and remained so even after he stopped giving injections. In fact they appeared to be permanently cured; but time and careful checking are required for assurance of this.

In this same group of functional bleeders are persons who bleed from the intestines and lungs, and I believe the venom treatment is the best so far discovered for that. Then there are persons who bleed excessively when a tooth is extracted or in an operation, and they have been helped by the treatment.

As for the other group of bleeders, the people whose blood shows actual changes or who have anatomical changes that account for their bleeding, the venom treatments check the bleeding but they have to be kept up perhaps indefinitely. It seems to be like the treatment of diabetes with insulin; there is a lack of something in the system.

I know that Dr. Peck has observed a number of cases of this type for more than three years, and they have done well under the venom treatment. Heretofore, the only alternative to treatment was removal of the spleen, and that is a pretty grave operation.

In this same group are people with enlarged veins and capillaries on the mucous membranes and they respond very well to venom too, but the injections have to be continued indefinitely.

I was interested to discover what effect these doses of snake poison have on normal people. It is barely noticeable, or not at all. Of course, the doses are extremely small—Dr. Peck I believe, gives them in doses that gradually increase up to one cubic centimeter, and the solution is one part of venom in 3,000 parts of diluting liquid. The only outwardly noticeable thing in bleeders is a black and blue spot where the injection is made.

After handling a batch of cobras, I was walking along the cages and looking them over. They had given up their poisons for a purpose far different from the one intended by Nature. The snakes were quiet now, and coiled, mostly in corners. Their lidless eyes stared in the glassy way of serpents. I thoughtfully looked at a big cobra that had turned halfway around between my gripped fingers. It was the second time he had done this. It had taken all my strength to restrain him. I remembered that in the Oriental branch of the Pasteur Institute at Bangkok, where poison was collected to produce serum for snake bites, cobras were chloroformed before they were handled. I doubted if they lived long under such treatment, but this was done in a country where cobras were readily obtained. With me it was a case of corresponding for months to obtain even a single cobra, hence my series represented considerable value.

"No," I was ruminating, "but I'll have to be careful with that big one."

I was pressing my thumb against my forefinger, and

idly looking at the two parallel lumps caused by the pressure.

A cheery voice spoke up behind me.

"Hurt your hand, doctor? Bit swollen, isn't it?"

It was Clyde Beatty who had dropped in to see me. His arena act at the circus in Madison Square Garden, with several dozen mixed lions and tigers, was making New Yorkers gasp. For bravery of the highest degree, and intuition in standing in that snarling mass of tan and stripes, and finally assembling the big cats in orderly rows upon their pedestals, he stands as the world's greatest animal trainer.

"Nothing the matter with my hand," I answered. "I've developed that muscle from gripping snakes like these. It's part of my job."

I asked Clyde how things were going at the show. He gave me details about some of the animals in his group that were trying to sneak behind him. I had seen that very thing several nights before, a lioness crouching, the muscles rippling for a leap, and Clyde Beatty as if warned by occult power, wheeling just in time to fire a blank cartridge and save himself. I would not have entered that cage for a fair-sized fortune.

And now he was discussing the training of the black panther, the most wicked beast on earth. His tone was calm. I was gripping the rail at the realization of such audacity.

"You can have your job, Clyde!" was my remark.

Before we started for a stroll among the animal buildings Beatty wanted to learn something about the cobras. I explained how the poison had been extracted, how the big cobra twisted to get free; how it would swing around, if it did get loose, imbed its fangs, and hold on.

"That's the one," I indicated, with a wave of my hand.

The motion caused the still irritable reptile to fly into a rearing pose. It flashed forward in a sweeping strike.

I heard an intake of breath.

"Doctor," said Clyde Beatty, "I'll return the compliment. You can have *your* job!"

CHAPTER XI

# RIOTING AMONG CELLS

In the question period following lectures, I am called upon from time to time to tell about research work with snake poisons. Frequently the questions have been about the development of snake-bite serum, but I am now being asked about the effects of poisons on baffling human maladies. These inquiries come up, of course, from word getting about regarding the investigations outlined in the preceding chapter. I must confess that answering such questions has often had my back to the wall, for a fair explanation to an audience not scientifically trained would be a lecture in itself.

Recently there arrived in my mail a booklet containing a lecture delivered by Sir Henry Gray before the Essex County Medical Society, at Windsor, Ontario, on December 6, 1933, and afterwards printed in the International Journal of Medicine and Surgery. It points to a solution of mysteries attending grave human ailments, and the action of remedial measures, among them the injection of the cobra poison. I wrote to Sir Henry and asked if I might discuss the lecture in this book, for most articles of the kind are read only by subscribers to the technical journals. He kindly granted permission. His article suggested a consideration of the actions of cells and the atmosphere they live in as being influenced by a system of government represented by the nerves. This latter is like combined federal, state and municipal governments. There are many departments, large and small. Sometimes the whole organization goes wrong, which means collapse. Again a local department may go awry, and in that area cause much commotion. Its insidious influence may spread.

The code of ethics among medical men engaged in human research is very exacting. Outside discussion of their work is quite generally forbidden until a formal paper summarizing results has been presented. In this way the studies of the physician are made public. By "public" I mean that the investigator is satisfied he has attained results of interest to the members of his profession, and that his checking of results is sufficiently definite to be convincing. There is a word from which he stands aloof. That word is "cure," and it is handled with the greatest caution. A line of research may produce highly beneficial results, and symptoms may apparently clear up among a number of patients; but only the cold facts are stated. Such patients are not declared cured in a formal paper, nor is a beneficial and newly discovered treatment referred to as a "cure." There may be recurrence. Results must be checked at definite periods following the publication of the paper. These periods may extend through months, or even years, before the originator of new methods in treating hitherto baffling ailments is satisfied to use the word "cure." The reason for this is to prevent the arousing of false hopes among unfortunate members of humanity.

Despite the neutral tone which necessarily attends these formal papers, many would be thrilling and fascinating to the public if they were translated into simple terms. I have read many and watched a number ultimately bring to the originator world-wide acclaim. Although I am not a medical man, my contact with medical men has made me familiar with technical wording, and so I trust I can fairly translate it into simple terms. That was what occurred in the preceding chapter, where I told about the work of Monae-Lesser and Peck. There again, ethics were duly considered. I could not have written that chapter unless both of the doctors had produced formal papers, even though I knew what they were accomplishing and had worked with them.

During the World War the surgical work of Sir Henry Gray at the front attracted much attention. One of his achievements was an operation he performed in saving the life of a soldier, whose heart was pierced by a bullet. His brilliant surgery afterwards gained him his knighthood. I first met him in the consulting room of

Dr. Monae-Lesser, during a discussion of the use of diluted cobra venom in the treatment of malignant growths.

In the giving of medication it is of prime interest to know just what the remedy does; whether or not it affects different organs, whether it changes certain of the body fluids, or neutralizes poisons in the blood. The search for answers to these questions is by no means generally successful. The immunizing of a horse by repeated, small injections of snake poison, until the blood serum is powerful enough to neutralize the venom in a human snake-bite victim, is a process not clearly understood. The laboratory worker knows what to do to produce antivenomous serum, he speaks about the development of "antibodies" in the blood of the immunized horse; but as to how these came into being and what they are, he is in the dark. A microscope shows no difference between the red and other blood cells of this horse and those from an animal which has received no treatment. That is the case with some other phases of medical research. The laboratory can repeatedly induce manifestations, without knowing exactly what takes place in the intricate organization of the body. Sir Henry discusses these mysteries in his lecture. It is entitled "Some Vagaries of Cancer and Deductions Therefrom," and starts by saying that it is the author's feeling that investigations of malignant growths have been focused too much on localized manifestations. His experience in the vagaries of the disease leads him to believe that those observers are on the right lines who think cancer is due to a change from the body's normal chemical control, or, what amounts to the same thing, loss of normal nerve control. This indicates a study of the body fluids, which take nourishment to different forms of cell life and tissues. Carried on at the same time should be the observation of abnormal areas of the body produced by misbehavior in the action of cells.

Under normal body control the various kinds of cells fulfill their parts in peace and accord. When such control is changed, local and general troubles become evident. Even slight changes appear to influence the activity of cells. This may be caused by a change in the balanced secretion of the body glands, or by diets lacking in certain forms of nourishment.

An example of perfect cell control is evident in the healthy person in the healing of a wound. Here is a mysterious influence regulating the multiplication of repair cells. This abruptly ceases when the repair has been effected. The scar, unless attending a severe laceration, is smooth, kept flush with the surface, and ultimately may be imperceptible. In other instances the cells continue at work after the repair has been completed. Their multiplication is not called off, and the scar tissue becomes raised and lumpy, forming a mass akin to malignancy.

While describing conditions which in this way produce misbehavior of cells, the article points to specific causes that may be throwing the internal forces out of balance. Diet, drugs, auto-intoxication, poisons developed in diseases, even mental influences may have internal effect. An example which appeared to belong in this last class was afforded by a patient from whom a malignant growth had been removed. Within a few months he was badly frightened to find swellings close to the area of operation and becoming larger. He returned to the operating doctor, who completely calmed his alarm and placed him under observation. There was little further enlargement. The condition remained passive for eleven years. Then, following an attack of influenza, the glands started to enlarge again, an operation was performed, and the condition was found to be malignant. In reviewing this case the article indicated the conviction that a changed mental attitude seemed the only new factor at work in altering the patient's bodily condition to arrest the growth. Finally the toxins left by the influenza attack upset the normal balance. The article states: "On another occasion I observed rapid increase of a tumor following an attack of influenza, the toxins of which seem to lower resistance in so many ways."

Another patient was troubled with symptoms indicating a small malignant area and was promptly operated upon. The mental condition preceding and following the operation was stubbornly gloomy. Although the affected area was small and despite treatment preceding the operation, the patient had lost much flesh. This appeared to be associated with a greatly depressed condition. The operation was of a nature to bring every hope of permanent relief. Following it, however, the condition of mind became even more depressed, and within a few months there was a recurrence of the condition.

With the citing of other instances, evidence accumulated to show that the metabolism (the process under which food is built up into living matter and the after balance of adjustment) should receive study as well as local malignant growths. Here the article calls attention to three methods of correction that may be used for altering the body fluids during conditions that seem to induce malignant growth. One relates to removing blood from a patient with an abnormal blood condition, washing it free of toxins and returning it to the circulation. This was first mentioned in the *British Medical Review,* "A Medical Review of Soviet Russia," and explains that Professor Spasokukotzsky of the Surgical Department of Moscow University had been trying a method based on the researches of Lintvarev of Saratov, who showed that in disease red blood cells contain a larger percentage of toxins than does the blood serum. Up to that time the professor had treated three cases of malignant growth by reinjecting the patients' washed red cells and the results were favorable.

Sir Henry explains that he treated three cases in this way, and all with definite benefits. His method was to withdraw fifty cubic centimeters of blood at a time. This is an amount that would about one quarter fill an ordinary drinking glass. The red cells were separated in a centrifuge, a vertical spinning shaft with hinged sockets for fitting long glass tubes. The device is turned so rapidly by a motor that the tubes swing outward horizontally, and the red cells of the blood are driven to the bottom. When the machine is stopped there are two strata, the concentrated red cells and the serum, the latter pale yellow. The serum is rejected and the red cells washed several times by mixing them with a salt solution corresponding to the salty content of blood. After the washing, fresh salt solution is mixed with the cells to make up the amount withdrawn and the cleansed red cells are injected into a vein. Injections are made every four to seven days according to the symptoms of the patient. In one case a dozen injections were made, which, according to the amounts treated, meant that in all about a quart of the patient's blood was extracted, washed, its original bulk made up by mixing with salt solution, and returned to the patient. In another case the injections were continued over a long period. With these cases, intensively watched, the results were astonishing in the extent of their success.

"Very little work has been done with respect to the rôle of red blood cells as controllers of disease," the article says. "The general idea seems to be that their sole function is that of carrying and distributing oxygen. In view of my experience, although my cases are far too few to make any dogmatic claims for it, I venture to recommend trial for this treatment. The side effects, so to speak, produced by it seem to be worthy of investigation and may prove to be of profound importance."

Continuing, the investigator reviews his use of cobra venom along with washing of the red blood cells. Here, again, he separates recognizable symptoms of benefit.

But to the lay mind the injection of a poison, even though it is diluted, would seem inconsistent with washing the blood from seeming impurities as outlined in Sir Henry Gray's paper. It should be understood, however, that some of the toxins that accumulate within the body, throwing the balance of health out of gear, are insidious and lurking. Slowly, but steadily they exert a weakening influence. They may be given off by those infinitely small organisms we call germs, by wastes that accumulate during prolonged fatigue, or appear during mental upset or worry where digestive processes are disturbed. A venom, however, like that of the cobra, is quickly neutralized after it has given the organization, as I heard one laboratory worker explain, "a kick in the shin." It is of an "intolerable" character. The way it weeds out evils affecting the organization of the nerves is astounding; and it is the atmosphere of that organization which influences the morale of various types of cells.

Cancer is a malady for which no germ has been discovered. But the biological chemist is shown to have an important part in solving the mystery, and his findings to be cooperative with the observations of the surgeon.

I sat for an hour in the Rockefeller Institute and watched motion pictures taken through a microscope by Dr. Heinz Rosenberger. These reels showed the activities of different kinds of cells. There were "good" cells and "bad" cells. In some instances the good cells had everything in order. While they behaved like miniature octopi, they seemed to be steadily housecleaning. Particles of débris were carted about, or swallowed and thus disposed of. Meandering cells that got in the way were pounced upon as vigorously as suspicious loafers in a police clean-up. Then, again, I saw mobs of cells gone wild and congregating in masses. The loyal types were outnumbered and unable to cope with them. They were hurled back from the rioting crowds. Coming from all directions were numbers swelling the mob, seemingly intent upon fusing their bodies into an abnormal mass, the beginning of malignant tissue. Thus, it seems, these minute forms can be sane or insane, sober or drunk, and those succumbing to the latter state may run wild and do crazy things. An old scar or chronic irritation attracts them. They swarm about it. The mysterious influence which starts their multiplication when a wound is to be repaired, is loosed, but now rages into a whirlwind of disorder resulting in the building of false tissue. This tissue breaks down because it is formed so fast that a supporting system of nerves and blood vessels cannot extend into it.

In the preceding chapter I briefly mentioned nerve centers and their influence upon the behavior of cells; but there are cells and cells. What I have been seeking is a simple description of the whole matter, and the most concise comes from a laboratory expert, who at times carries tubes of germs around the place in his inside vest pocket to give them a good "cultural" temperature, when the bacteriological ovens are overcrowded. I submitted his definition to Dr. Monae-Lesser, who, with a smile, said it was "simpler" than he would have thought of expressing it, though outlining theories now substantiated.

"Tommy," I said, "you know Monae-Lesser's work and you've read Sir Henry's paper. You live in an atmosphere of micro-organisms, cellular and intracellular tissue, but you've got a knack of talking about these things in a plain way, like calling them by their first names. Give me a definition in those terms, of what can happen to a patient with misbehaving cells when cobra poison is injected."

"Figure it like this," said Tommy. "The maintenance of an animal body is like the running of a city. It's complicated. Nature is sometimes caught napping, and cell discipline, which means a lot, may slip. If cells are inclined to be unruly and the home atmosphere gets worse, they go on a tear, and still Nature is inclined to be tolerant. The medical investigator notes this. Something has gone wrong and he seeks to correct it. He may be fairly successful. Again, there are measures that may be more successful. According to the snake poison theory you have a new line of attack to regain the normal or start Nature to investigate, to call in reserve forces that have not been utilized, and set the patient on her or his feet."

"What forces?"

"That's beyond me. Who can define electricity?"

"But blood washing, to take out a poison; then the injection of cobra poison?"

"Analyze Monae-Lesser's deductions. They're ultra-technical, but point to abnormal excitement of nerve centers. Toxins in the blood excite the nerves and communicate the excitement to the cells, which are builders of tissue. Absorbing propaganda of the worst kind, the mobs keep coming in contact with bad nerve centers, rush out on wild parties and do fool things."

"But the cobra poison—"

"I'm coming to that. It's of the paralyzing or numbing type. It puts the nerves to sleep, and Nature, at last concerned with their mumblings during the coma, calls in a 'Force.' I might say that she finally sends for her family physician, whose ministrations are secret and mysterious, and the nerves come out from the treatment, much improved. Each time they are put to sleep they receive further treatment, so they get well, stop their foolish ranting to the cells, and the party's over—and much to the relief of the white cells, or leucocytes, which have had as hard a time as a handful of cops trying to control a mass of crazy rioters.

"That's the difference between your neurotoxic or nerve-numbing poison of the cobra injected into the body and the toxins of disease, which only tend to slowly sicken the nerves."

That was Tommy's way of describing the results of complex research in plain terms.

CHAPTER XII

# INSIDE HISTORIES OF PETS

"ALWAYS those devilish snakes and their poisons," remarked a member of the Zoological Society. "Why don't you work up some research with the big animals?"

He continued his friendly criticism:

"I think you are missing a big opportunity in not undertaking a study of animal psychology. You could make close friends of dozens of your animals."

Here was a point requiring a tactful answer. My visitor was a well-to-do man, fond of dogs and keeping several horses. He had his favorite animals in the Park, and always brought a bag of cakes and some lumps of sugar. His favorites were the kinds the keepers called "grafting" animals—those alertly friendly when a visitor indicates that something edible may be handed out. My critic was deceived by them. He was accustomed to the ways of disciplined, domestic stock.

I told him that my job with the mammals was to keep them as contented as possible in their enclosures, watch their feeding and maintain contact with the keepers. Then there were over five hundred labels to be kept up to date in harmony with the latest usage of scientific classification. Scratching claws or fading paint would make many of them illegible if they were not constantly inspected and repaired. Also there were hundreds of locks for which I was responsible, and these must be watched.

My explanation was quite lengthy, and I think my friend gained a new insight into the job of maintaining an animal collection. I told him that the scientific investigation of mammals largely fell to the department of veterinary research, which made studies of animal parasites and abnormal conditions, conducted post mortems and worked with the microscope. My own observations of animal psychology had been applied in the direction of improving our methods of keeping the animals thriving. But as to pets, I didn't believe in playing favorites; and when some of the keepers had been so inclined, there usually had been trouble. One reason for this was that some animals, with which keepers were particularly friendly, ceased to fear the men, and should they be of dangerous types were inclined, if they lost their tempers, to turn on them. Of course there were exceptional cases. In every collection of animals there are a few that stand out as being abnormally gentle, and so become natural pets.

I do not mean to suggest that I am against the keeping of pet animals. I am always sympathetic towards a child who is keen about pets—provided the interest runs to harmless and suitable kinds. But there are pets and *pets*. I have had years of observing owners' reactions to them, some of them quaint or pathetic, others exciting or highly humorous.

I have had disturbing experiences resulting from owners deciding that they must dispense with their pets. The most frequent disposals are of monkeys that have become ugly, but are still regarded with great affection.

Such owners see one of the big cages in the monkey house undergoing repair, and empty. They at once think that here are ideal quarters for "Mike" or "Ginger" or whatever the animal's name may be. He can live *alone*, and so not be bothered by the other monkeys. I explain, at first in vain, that the simian making faces at me from the box or basket in which it has been brought, cannot be given a big cage by itself. The owner may even leave in a huff. But after the monkey has ripped down some lace curtains or bitten the cook, it shows up again.

This time the owner consents that it be placed in a cage with others of its kind and arguments seem to be over. I am writing down the name and address of the donor, when the lady leaves the office saying she will be back in a moment.

She returns from her automobile with a doll's bureau in which there are several drawers, and says that the monkey is fond of playing with this, and that it must be placed in the cage or he will be homesick.

Here is a problem. I do not like to tell the lady that there are rogues in the cage who would take that bureau up to the very top and drop it on the heads of their associates, quite possibly initiating her own precious pet by such a process. The bureau is accepted, but quietly stowed away among other impractical toys that have thus accumulated.

The presence of that monkey is liable to entail tilts with the owner. She may come to my office a week later with fire in her eye, and complain that there is a bossy monkey in the crowd with her pet, and that the bully must be taken out.

There is seldom a cage of monkeys in which one or two are not inclined to claim leadership. This results in some screeching and chattering at feeding time, but ultimately all the members receive their share of rations. Injuries from fighting are rare among monkeys. As to natural bosses in a flock, I have seen this condition among troops of wild monkeys in jungle areas. Such explanations do not calm the former owner of the pet monkey, so I get ready to dodge when I see her coming, as I am dodging others who have built up similar troubles.

One lady presented a monkey that happily *could* be placed alone in a moderate-sized cage that was empty. It looked as if there would be no adjustments about this pet. But the owner returned and asked the keepers to place a chair in the cage, and there she sat day after day to talk to her pet. The spectacle of the lady in the cage caused a jam of visitors in that part of the building.

Next to monkeys, raccoons and squirrels are most often presented to us, and with these there is frequently insistence upon the installation of home-built exercising devices. A pet squirrel is brought in, and by tactful questions I manage to discover the reason for its presentation. It has gnawed a hole through the upholstery of a couch. The squirrel couldn't be blamed, as the heavily stuffed covering looked like an ideal place to make a nest. The owner insists upon the installation of an exercising wheel. This is brought in from an automobile in sections. It is a yard high, has a decorative gilded stand, and attached to the wheel is a circlet of sleigh bells. I state firmly that the device will frighten the wits out of all the other squirrels in the exhibition cage, or at least visitors will complain to that effect. The owner is peeved, although we already have seven squirrels like her pet, brought up from infancy together and living as a peaceful group. That is one reason for donated pets being disturbing. They duplicate certain species and are liable to start fighting among previously quiet animals. I consider that half a dozen raccoons are ample for exhibition, but there was a time when donations brought the number up to twenty-two specimens.

Owing to years of such difficulties we recently formed a decision about the acceptance of pets. If they were of types where the exhibition quarters already contained the allotted number, the condition was explained to the owner so that refusal of acceptance would not be taken amiss. Even so this doesn't work out smoothly in all cases.

Lee Crandall, the Park's curator of birds, tells me that his troubles have been worse than my problems with the four-legged animals. His main problem has been with canaries. He acquired so many pet canaries that they swarmed as a flock. Barely a day went by without someone standing in front of the big panels and calling "Goldie" or "Dickey" and waiting for a yellow form to flutter out of the legion for recognition. When nothing of the kind happened there was a search for the keepers to ascertain if the pet had died. The keepers were distracted in seeking to establish individuality among the canaries, and in some instances irritated former owners by pointing out the "wrong" bird. It got so difficult and confusing that the acceptance of canaries was banned.

The bird curator's next most serious problem was with parrots, for would-be donors constantly asked him to furnish cages for sole occupancy. He acquired so many parrots that a ban was placed on them and keepers received instructions not to accept another parrot.

One day the curator came into the building to find his head keeper looking very worried. Before him was a cage containing a parrot, and bending over it was a lady in tears. While this was an ordinary green parrot, it was accepted, as the owner's story was unusual. Her husband, who had died a week before, had been very fond of the parrot, and it habitually began calling the man's name late in the day, at the time of his return from business. The parrot kept it up, day by day. It would call, repeatedly, and turn its head waiting for an answer. The widow was unable to endure it.

Occasionally pets of an exceptional nature are brought into the Park. I had a telephone call from a lady who wanted an appointment to talk about a turtle. The matter seemed trivial, but I told her what time I was around the office and forgot about it. She was later ushered in, carrying a small pasteboard box, and said she had come to give me the turtle. With this, she drew some sheets of closely written paper from her handbag, and I settled back for what looked like a boring discussion. Within five minutes I changed my mind and was intensely interested, although the specimen, now disclosed, was only of a common kind of spotted pond turtle.

This turtle had lived in the kitchen of a farm house in White Plains for forty-two years. During this period it had been the pet of the lady's grandfather and her father. Its quarters had consisted of a large open-top box, with a pan of water at one end. At times it had wandered around the kitchen where it had a favorite bunk behind a woodburning range. Its claws were grotesquely long and its upper jaw, or mandible, had grown to such length that it looked like a beak. Placed on a corner of my desk it would return to its owner and snuggle against her hand. The old house was to be torn down, she said, to make way for improvements, and the family was moving into New York. She thought the turtle would be happier in the Park than in an apartment.

Another turtle, a box turtle, was brought in by a family moving to Minnesota, where they thought the climate would be too cold for their pet. This turtle had lived in a back yard in the Bronx, and wintered in a basement, for thirty-three years. Another box turtle came in under similar conditions; it had been a pet for twenty-seven years. Both of these box turtles were full grown when first coming under observation, so an estimate of their age was problematical.

From correspondence seeking remedies for the maladies of turtles, it would seem that many people have such creatures as pets. One of the principal troubles of captive turtles is an irritation of the eyelids, causing the lids to swell and remain closed. The trouble is remedied by giving the reptile a bath in mildly saline water. I have received letters of the warmest thanks in appreciation of advice which had resulted in bringing a pet back to good health.

There are quite a few pet snakes in the country, but the most extraordinary example of interest in serpents which has come to my attention concerns a high-standing and veteran member of the Federal Government in Washington, who does not profess to be a scientist. His hobby is rearing pythons, one at a time; and with his exacting care they grow like weeds. I have known him over thirty years and during that time he has reared five pythons and one anaconda. After five years the snake attains such a size that it outgrows his apartment and he presents it to some zoological institution where he is assured it will get the best of care. Then he starts rearing another infant serpent.

The outside interest in snakes has caused two big stirs in the reptile house of the Park. One was the attempted theft of two cobras, when the thieves were frightened away after nearly cutting through the steel door of the cage of these deadly reptiles, and the other was the successful theft of ten of the most beautifully colored snakes in the collection, all of harmless kinds.

The entrance, in the latter instance, had been made by forcing a window. The small snake cases are set in columns, in receptacles which keep the sliding glass fronts from moving. They are not locked, as is the case with all the larger and poisonous snakes. The cases were found to have been slid outward, the glasses had been opened and the following specimens were gone:

A Brazilian red-banded snake, a spotted racer from

Colima, Mexico, a rainbow boa from Venezuela, an infant Brazilian boa constrictor, a Bahama boa, an Arizona desert racer, a Florida corn snake, a red-bellied water snake, a California king snake and an indigo or gopher snake from Florida. The indigo snake was much the largest, being about five feet long. Also it was the least valuable. The rainbow boa had been in the collection for sixteen years. The Brazilian red-banded snake was a personal gift from Dr. Vital Brazil, and had been given to me while I was visiting him at São Paulo, at the serum institute, ten years before. Moreover, this series of attractive reptiles had been arranged as illustrations for biology classes from the schools, who visit the Park for instruction and lectures.

The detective bureau was given details and the newspapers published stories describing the colors of the missing snakes. The Brazilian snake, with its alternate bands of crimson and pale yellow, was characteristic. So was the Mexican racer, looking as if each scale was set with a small emerald. In fact each and every specimen had individuality and was rare, except the big blue-black gopher snake from Florida.

The very next day Detective William O'Brien picked up a scent. A snake had been killed in Harlem. It had appeared in an apartment in the negro section as the head of the family arose to touch a match under the morning coffeepot. The snake reared into view in the kitchen, and so thoroughly was it dispatched that the handle of a broom was splintered. Wrapped in newspaper, the battered serpent was taken to the nearest police station. The lieutenant at the desk was unable to identify it, beyond deciding that he didn't like its looks, and it was consigned to an ash can. But Detective O'Brien, miles away in the Bronx, heard of the incident. He picked up Keeper Frederick Taggart of my staff and sallied forth to see if the snake was one of the missing ten.

It turned out to be a harmless Texas bull snake, but O'Brien was not satisfied. He wanted to know where that snake came from and located a tenant in the house who had some snakes. The tenant's premises were subjected to a careful search. Two separate nests of serpents were discovered in battered satchels. One consisted of an assortment of non-venomous Texas rat snakes, with a few bull snakes mixed in; the other caused Taggart to stiffen a bit as two big rattlers glared up at him. The owner explained that his menagerie was used to encourage the sale of rattlesnake oil.

The detective gleaned from this itinerant snake fancier that another man who had much to do with snakes lived in a cold-water apartment not far away. O'Brien and Taggart went to the given address.

They found a man surrounded with snakes of various shapes and colors. But they were made of springs covered with paper. The trail was lost right there.

Later I had a talk with Detective O'Brien. There was one point which bothered us. Nine of the snakes could be positively identified as ours—if we saw any one of them. All were rare, and I doubted if they were duplicated in collections. All but one were small. The difficulty was about the indigo or gopher snake. That was a kind often carried in small shows; even by the "medicine man" in Harlem. Gopher snakes are cannibals. If the thief had carelessly placed all the stolen snakes in one receptacle, the gopher snake had swallowed the others and hence left no means of identification. The detective scratched the back of his head with a pencil.

"You've handed us a new line," he said.

A few days later I received a telephone call from the assistant principal of one of New York's high schools. One of his teachers had read the newspaper stories of the disappearance of the snakes, and had doubts about two of her pupils. They had brought three snakes to the biology room, saying that they had just captured them in New Jersey. As the temperature was hovering around zero, the capture of snakes in such weather seemed extraordinary. I went to the school and identified two of the snakes. They were the Bahama boa and the California king snake. The third specimen was a big gopher snake, which undoubtedly was ours. All were in separate cages.

Through the smart work of the assistant principal

and Detective O'Brien, two boys, pupils of the school and exceptionally keen in biology studies, confessed to taking the serpents. They disclosed the rest of the snakes hidden in a suitcase, in a closet of the school's biological department. The boys said that it was their interest in reptiles and fascination by the rare and prettily colored specimens which led them to break into a public building after closing hours and take the snakes. Investigation disclosed that they had already made close study of the specimens and prepared loose-leaf filing cards, noting details of coloration, numbers of scale rows and precise lengths. One of the cards bore a notation about success in inducing one of the snakes to feed. The boys were fifteen and sixteen years of age respectively, and their youthful fanaticism had obscured the realization of their misdemeanor. But the matter was so serious that a formal complaint was filed in the children's court. The disciplinary measures were stern, but because the case was so remarkable they were tempered with mercy. Newspaper articles assured the public that the missing serpents were again behind their plate-glass exhibition panels.

Another case of intense interest in a serpent occurred in the downtown artist section of New York. It came to my attention when an artist visited me for advice about maintaining a fifteen-foot anaconda. As these big, lazy constrictors of the American tropics are never found far from water, I suggested that the snake should have a bathing tank, in which it would lie two-thirds of its time.

It seemed that the artist was to decorate a fountain or something of the kind with a big serpent in carved stone—hence his purchase of the anaconda. He said the loft he had taken had plenty of room. I told him how to build a pen about twelve feet long and in one end of it have a wooden tank about five feet in length, thirty inches high and a yard wide. The tank would provide ample quarters for the serpent, and the dry area a place for it to spread out while acting as a model. For food I suggested one unplucked chicken a week. This could be obtained at any of the markets.

There was no further news about the anaconda for two months, and then the artist blew in again. For the moment I couldn't recall who the man was, but he brought the whole matter back by saying that his anaconda was in superb condition, eating a chicken a week and luxuriating in its tank. It had taken a long time to complete a big clay model of the serpent, as the anaconda spent so much time in the water; but he had been patient and had grown very fond of it. He thought that the iridescent bloom of its skin was a reward for giving it ample chance to bathe. He intended to keep the serpent and later take it with him to Europe, where he had further work to do. Now he was going on a vacation and had offered another artist the shelter of his studio. This other artist was broke, but had some jobs in mind. He didn't object to living with an anaconda, and had been provided with funds to buy unplucked chickens.

The next chapter in the anaconda's history was ushered in three weeks later. The owner rushed in upon me. He was greatly agitated. I was prepared to hear a tale about his returning to the studio and finding a man squeezed to death. But there was an altogether different story.

The snake's owner had returned a week earlier than he expected, and dashing up the stairs banged on the door. The other artist was out, but the rightful occupant had a duplicate key. He strode over to inspect the anaconda's quarters. The serpent was on the dry side. But what was this!

The netting over that part of the enclosure containing the tank had been removed; and across the enclosure was a fence of boards. In fact the snake was confined—but to the *dry* part of the enclosure.

The most startling discovery, however, was the milky appearance of the water in the tank, a floating cake of soap and a towel thrown over a nearby chair. So this was what had happened! A fellow artist, given kindly shelter, had chased the anaconda from its tank, and was using it as a bathtub!

I broke into the story to ask what had happened when the other artist returned, but my informant, stuttering with wrath, said he hadn't waited for him. He had rushed up to me to get some balm or lotion for the anaconda's scaly hide, which looked dry and pitted. I told him such treatment was unnecessary. He should get rid of the soapy water and provide fresh, after which he could chaperon his snake back to its tank. As he prepared to dash away I thought it well to suggest another point, and that was his *personal* tendering of a chicken to the pet snake, at once.

## CHAPTER XIII

# THE VAMPIRE REAPPEARS

My assistant and I were engaged in a final discussion of the vampire paper.

"What!" I expect the reader will mutter. "That paper not ready yet—and all it concerns is one bat!"

But that little bat, which brought the assertion from Head Keeper Toomey "that it had worked hard for its keep," has furnished a fine example of how a scientist can build up his hopes, have the props knocked from under him, get up and brush himself off, and go at it again to find the struggle well worth while.

My young assistant has been accorded the part of junior author on the paper. It is the first lengthy scientific treatise on which his name has appeared, and naturally he is proud of this. He has worked long and hard searching the libraries of his university for all literature available on the vampire bat. I have been doing some of this work in New York, looking into different kinds of books from those I have asked him to examine.

At last we are ready to type the paper. We propose to discuss the origin of the term "vampire," proceed to the time when early scientists really discovered a sanguineous bat in the New World, and describe their tilts regarding its classification. Thence the article will swing into the effect of the discovery of a bat living upon blood, upon the superstitious legends of Europe, tell how the vampire bat's habits were variously described by early naturalists, how natural history books called it and its near allies "blood-sucking bats," and how they were all wrong in that assertion. The habits of the vampire, as observed at the Park, had apparently never appeared in print. We were discussing all this, based on our search of the literature.

"Well," I said, "it looks as if all the natural history books were wrong in calling the members of the *Desmodontidae* * 'blood-sucking bats.' They don't suck blood, but lap it with the tongue. Do we go on record in claiming priority in describing that?"

"No. Dunn, in the *Journal of Preventive Medicine* in 1932, says the vampire laps and does not suck blood."

"Huh!" was my response. "There goes a star point. Who will see Dunn's description except medical subscribers to the Journal—and few of them are interested in vampires. O.K. We'll quote and credit Dunn and throw in what I thought were new observations. How about the quadrupedal gait I studied?"

"I found an old reference intimating this. The Reverend J. G. Wood mentions that vampires can

* The vampire bat family.

LAIR OF THE SPEAR–NOSED BAT

Ruins in Central America, centuries-old, with deep and dark galleries orm the homes of many tropical bats.

SPEAR–NOSED BAT

These sinister-looking creatures, with wing-spread up to a yard, were lleged by early naturalists to be vampires, but they feed largely upon fruit.

ADULT AND INFANT VAMPIRE

The baby is five weeks old. The point to be noted is the freedom of the vampire's limb from wing membrane, enabling this creature to *walk.*

SPEAR–NOSED BAT

Note the attachment of wing membrane across the limbs, which is to be generally seen among bats except the vampire.

walk, rather than grovel like other bats. But the description is insufficient to indicate that it is their habit."

"Yes," I explained. "Wood brought out the idea that there was something unusual in the way the bat moved about other than in flight. Beebe felt one crawling over a sheet in the dark, and described its gait as sort of a creeping walk. But he said 'I hardly felt the pushing of the feet and pulling of the thumbs as it crawled along'—and that's not what really happens. The vampire stalks with folded wings, like a slender, four-legged animal. Its long 'thumbs' are turned out at right angles and serve as *planted*, front feet on the wing stalks."

"That seems to be an original observation—so far as written description goes."

"And the vampire leaping upward from the ground, to launch itself into flight?" I queried.

"Haven't seen that," he answered.

"The birth of a vampire, its rapid growth and drinking blood of its own accord within three weeks?"

"That's new. And your motion pictures appear to be the only photographed records of the combined habits."

So the review at last goes into type after my assistant has gone through two hundred and twenty-two books and scientific papers and I have read through about fifty others, two hundred and seventy-seven publications examined all told, besides nights of watching the living animal. All started by one little bat, and resulting in a review that will change future descriptions in natural history books. The two main points brought out show that the vampire is a blood-lapping, not a blood-sucking creature; and that it stalks about with its body reared from the ground like a spider, with such a feathery gait that it can creep over the bodies of large animals and human beings, looking for a favorable place to make its painless gash with lance-like incisors, without waking its victims.

During the nights I spent with my captive from the Chilibrillo Caves, I had been enthused with the thought that habits not previously known were being watched; yet Dunn had already formally brought out the blood-lapping habit, and others had been close to solving the walking gait. This goes to show the caution necessary in making claims of priority when writing a scientific paper.

That paper brings out some highly interesting points. The term "vampire" originated long before the discovery or any thought of a so-called blood-sucking bat. In later years, however, the actual discovery of the bat appears to have added elaboration to the vampire tradition. The word "vampire" can be definitely traced to Slavonic sources. It was first applied in eastern Europe to alleged blood-sucking, supernatural beings, supposed to be the souls of dead persons, which left the interred body at night in one of many forms, to suck the blood of sleeping persons, and sometimes of animals. Of the numerous shapes then said to be assumed by a vampire, it is to be noted that there were no early references to the bat-shape. These came into being much later. The early forms associated with vampirism were often uncouth monstrosities, the werewolf, dogs, cats, birds of various kinds, snakes and even such inanimate guises as long straws or floating flames. Actually, superstition about blood-sucking creatures has been widespread and of dateless origin. It was known in many of the ancient cultures of the Old World.

It existed independently among the Mayans before the arrival of Cortez in the early sixteenth century. The presence of the tropical American sanguineous bats, which ranged plentifully through the Mayan country, provided a foundation for the superstition. Here was reverence for a blood-sucking bat god.*

The return of the followers of Cortez to Europe with tales of blood-sucking bats, founded on definite knowledge of such creatures, may have strengthened the European superstitions concerning vampirism. From a chronological examination of the old literature it seems that not long after the return of the Spaniards there were allegations about blood-sucking bats in Europe, where no sanguineous bats have ever occurred. Curiously enough, the "discovery" of these creatures did not

* *Mythology of All Races,* 1930, Vol. XI, p. 177, Archeological Institute of America.

appear to become associated with legends about the vampire. Bats had long had an eerie reputation, fitting in with tales of witchcraft and the like, and the stories from the Cortez expeditions furnished fine support. But the bats were not for a time thrown into the wild hodge-podge of forms, ranging from werewolfs to flames, assumed by "vampires" as they saw fit.

However, it was not many years after the conquest of the New World tropics that a vampire scare developed in Europe. In 1720, particularly in the Slavonic countries, all sorts of works, scientific and philosophical, appeared relating incidents and cases of unfortunate people who became afflicted with vampirism. Slowly the supernatural "vampire" acquired an added list of assumed forms. But modern times were approaching, and adventurous travellers were returning from tropical America with tales of blood-sucking bats, both the size of the bats and their blood-thirstiness well exaggerated, when the European "vampire," up to this aloof from bats, assumed wings on some of its rampages. Finally there appeared the remarkable novel "Dracula," printed first in 1897, in which the bat form plays a prominent part.*

The early scientist setting out to explore Central and South America knew there were bats which fed upon blood, for stories were rife among the early set-

* *Dracula,* by Bram Stoker. Doubleday, Doran & Co., Inc.

tlers. But the exact kind of bat was unknown. The Old World term "vampire" was just the thing to apply to the creature, and the title of Vampire Bat came into being. The ugliest and largest bats were placed under suspicion. Some of the early observers were not trained like our modern scientists. Imagination was drawn upon to describe the habits of the big, spear-nosed bats, which are largely fruit eaters and not blood drinkers. One of these attains a wing spread of a yard. It was said to hover over sleeping victims, fanning them with its wings to induce sleep before it lanced a vein with keen teeth, then inserted a slender tongue and sucked man or beast dry. In fact, the great naturalist Linnaeus, in 1766, gave this big but harmless bat the scientific name of *Vampyrus spectrum*, and owing to the rule of scientific precedence in the placing of a name, it will remain thus libeled.*

It appears that the actual vampire bat was discovered as a species, given a scientific name and placed in the classified lists before its sanguineous habits were known. Prince Maximilian zu Wied, a noted scientist and explorer, made for it in 1826 a new genus, *Desmodus*, thus separating it from the group of larger, spear-nosed bats into which preceding studies had dropped it. By 1839 its blood-thirsty habits were established, and another naturalist, Waterhouse, referred to it as *the*

* Another similar instance among names based on inferred habits is that of the common blacksnake, named by Linnaeus *Coluber constrictor*. The blacksnake is not a constrictor.

Vampire. In recent years it has been recognized as being different from other bats on account of its lancing teeth, and has been accorded a place in a *family* practically its own. Only two other small bats are associated with it, these representing the genera *Diphylla* and *Diaemus*, each with a single species smaller than the vampire, but undoubtedly sanguineous, to judge from the form of their lance-like incisors.

But to revert to the notes of early naturalists made since the vampire bat had attained the dignity of a scientific name. All descriptions brand it as a blood-sucking bat. This is indicated in the writings of Androvandi, Shaw, Cuvier, Buffon, Geoffroy Saint-Hilaire, Swainson, Gervais, Hensel, Goeldi, Quelch and others. Even the immortal Darwin was inclined so to believe. Investigation of the American tropics went on, and scientific lists were growing rapidly with the addition of new species of mammals, birds, reptiles and amphibians. Observations of remarkable habits of many kinds appeared in print, and many of these have since been re-checked as quite correct. But the vampire continued to fool the eminent explorers. Their horses and even the human members of their parties met the dawn streaked and stained with blood, or with wounds still oozing from bites and blood-letting, painlessly administered. Vampires were seen to flit around the camp fires and to be at the horses; but those were the days before the electric flashlamp. A blazing faggot from the fire, used as an investigating medium, sent vampires scurrying into the shadows.

Strangely enough, in spite of modern methods of investigation, observations failed to appear in print to correct the wrong impression. Several modern writers have gone into details explaining how the vampire applies its lips to the wound, and in my library is one of the latest and most pretentious sets of natural history volumes, containing a heading "The Blood-sucking Vampires."

Dunn's formal paper in 1932, found during the search for references, seems to be the first description of the real habits of the vampire when feeding. My observations made during weeks of watching the bat come to her dish of blood and quickly lap it up, were only a checking of Dunn's description with another specimen.

When I had watched the vampire at night for three months, and not once seen her purse her lips but always employ the darting tongue, the disturbing thought occurred that mine was only a captive animal and drinking from a dish. Her actions might be different from those of a wild bat. That was a real problem. While it was on my mind a fortunate thing happened.

Paul Bransom, the noted animal artist, whom I have known for years, came into my office with a tall, bearded stranger. At the first flash I decided I was going to like this man. His eyes impressed me. They were blue, quiet but intense, and of scintillating clearness. He was Sacha Siemel, a professional jaguar hunter of Brazil, whom Julian Duguid met during a wilderness expedition and afterwards in his writings called "Tiger Man." The title came from Siemel's work in tracking down jaguars that had been destroying cattle. The big cat is known as the tiger all through its tropical range, or to indicate it in Spanish, *el tigre.*

Siemel is a man of fine education and a born naturalist. Moreover, he weighs his words. I asked him if he had ever got close enough to see wild vampires feeding.

"Yes," said Siemel, "but they do not suck blood. They lap it."

This made me sit up and glow with gratification. The explorer explained that he had made the observation in an area of southerly Brazil, close to the Bolivian frontier, a region of swampy forests, while conducting a party using a number of horses. The expedition was seriously bothered by vampires attacking the animals at night. Several times, as Siemel inspected the tethered animals with a flashlight, he saw bats clinging to their shoulders, the bites having already been made. The bats were bold enough to remain on the horses while he approached to within a few feet and watched for two or three minutes, before they flitted into the darkness. The examinations were long enough to see that all the bats were rapidly lapping the blood with their tongues from the edges of the wounds.

"And they *crawled* over the horses' backs, like four-legged things," added Siemel.

This made me sit up again. More verification, but showing that I hadn't been the first to study it. However, it bolstered up a theoretical assertion in my notes that "this quadrupedal gait is probably utilized in lightly stalking over the back of a sleeping animal in seeking a spot to use lancing-type incisors."

It seems improbable that so many bats can expect a blood feast every night, when no human beings, horses or cattle are present to disturb the quiet of the forest. Some of them may; for deer and tapirs roam the forests, and even a gorged jaguar might slumber through a session of blood-letting. From the way I have seen the larger monkeys sleep, I'd be inclined to believe that the spidery stalk of a vampire could get away with it among some of the simians. I think, though, that there can be little doubt that vampires also hunt and devour small rodents and reptiles. But when a group of big animals shows up, a flock of vampires forgets all else in its blood-lust.

Conversations with persons in the tropics who have been bitten by vampires, and the notes of reliable explorers bitten while sleeping in their camps, all tell the same story—the discovery that the thing has happened comes in the morning, upon normal waking. There is usually a considerable show of blood that has continued to flow from a very small wound, *painlessly* produced. Every member of an expedition from a western university, reconnoitering in Colombia, was bitten during a night in camp. The half-dozen members awoke to a gory spectacle. Some had been bitten by several bats, yet not a member of the party had felt a trace of the bats at work. This ability to lance the flesh quickly is shown in an accompanying photograph of the bat's teeth. The upper incisors are shaped like a razor-edged scoop. Here, apparently, is Nature's last word in providing an instrument to produce a cut that will bleed, and may account for the long-continued bleeding. Opinion seems to be divided as to whether the saliva of the bat contains an anticoagulant. I am inclined to think so, but it will take more than one more specimen to enable me to carry out such an investigation. Perhaps it is fortunate that, since the death of the specimen which got me started on the problem, visitors at the Park demand that we replace it. And that makes a definite excuse to get to the tropics again! Incidentally if I do get another for exhibition, I'm looking for a scolding from a certain editorial writer. He said I shouldn't have brought back the first vampire, that it was a gruesome thing and might have escaped. Possibly the bat publicity lured him to read "Dracula."

Curiously enough, throughout the vast area of the Old World tropics—of Asia, Malaya and Africa, where there is a great variety of bats large and small, some attaining a wing spread of five feet, the big fellows fruit-eaters, no blood-drinking bats are known to occur anywhere. In a group called "false vampires" there is one species, *Megaderma lyra*, known to capture and devour birds, other bats and even pounce on mice. In these habits it resembles the spear-nosed bat, unique in the American tropics in being as savage as a weasel in killing small animals. There is another Old World tropical species which skims over lagoons and rivers and catches small fish. It seems incongruous, however, that with the superstitions of blood-sucking originating in Europe, the only strictly sanguineous mammals known should be confined to the New World, and have applied to them a term growing out of Old World superstitions.

CHAPTER XIV

# MONSTERS OF A YEAR

It was Monday morning and everything was placid in my office. There were no reported fights, temperamental antics or illness among the members of the big animal family surrounding me. Scientific jobs had been cleaned up for the moment. An orderly pile of mail lay before me. It looked like a smooth start to the week, with no problems in sight.

The telephone rang. After a moment, my secretary swung the extension arm over to my side of the double desk.

It was the Associated Press. A big and mysterious animal had three times been reported in the waters of Alaska. What might it be? Was it a sea serpent?

I said that I didn't believe in the kind of sea serpents periodically alleged to make their appearance, that science failed to concede their existence as no body or bones had ever been washed up on a beach, and that no such creature had ever stopped a harpoon or been shot.

But the A.P. insisted that reports from masters of three separated vessels meant that *something* had been seen. Was there a possibility that a few prehistoric monsters still lived in the sea?

I insisted that the chance of a prehistoric monster remaining alive was extremely remote. I could not, of course, declare that the ocean was free of them. A short article appeared among the newspapers discussing the "monster."

It was not long before additional reports came in. The creature had been seen as far south along the Pacific coast as Oregon. The "monster" gathered fame. One enterprising owner of a vessel suggested taking sight-seeing parties out on a cruise in search of it.

Following the tales of the Pacific monster, Scotland furnished the next similar sensation in the appearance of a large and mysterious creature in Loch Ness. Here was a far more convincing manifestation. Dozens of people saw the creature, and many made drawings of it. With the monster popping into sight week after week in the waters of Loch Ness, yet another story thrilled the readers of the world's newspapers. It related to the body of a strange, marine creature, twenty-five feet long and some eighteen feet in circumference, washed up on the beach at Querqueville, near Cherbourg, in France. The creature had a thin neck, three feet long, and its head resembled that of a camel.

My reputation has been much associated with reptiles, and marine "monsters," to the press, seemed to indicate reptilian relationship; so the epidemic of scattered mysteries gave me plenty of work in giving interviews and answering questions by mail and telephone. Starting on a placid Monday morning, the problem gathered weight like a snowball rolling down hill.

That there is some foundation for the sea-serpent-marine-monster belief is evident from what has occurred during 1933–34 and in the past, although the nature of the creature off the Pacific coast, and strangely enough, of the much-observed Loch Ness "monster" remains undetermined.

From the many tales I have heard and the deductions that have been made, I think that the belief is a thought too thrilling and fascinating to be much weakened by scientific discussion. The oceans cover about three fourths of the world's surface. There is thus the sense that a ship at sea is making its way over a vast surface under which lurk varied mysteries. In parts the ocean is more than five miles deep. Oceanographers state that the bottom has its mountain ranges, with their ledges, caves and canyons. There are stretches which parallel the deserts of the land, there are great jungles of marine vegetation, through which prowl creatures more savage than the tiger, creatures known to science. But to the lay mind the thought of these great depths suggests lairs of survivors from the age of prehistoric monsters, creatures which may occasionally come from these depths to be glimpsed by the lookout of a passing vessel.

When such a suggestion is presented to the scientist, he brushes it aside with the declaration that in the depth the pressure of the water is tremendous. There are marine creatures which man has laboriously dredged into view. They are adapted to that pressure, but they must remain within it. Their compressed internal organs are literally torn apart when brought to the upper strata. If a sea monster should rush from great depths to the surface, it is argued that parts of it would really explode. Hence, if prehistoric monsters still exist in the oceans' greatest depths, they must stay there, and man may never see them, unless the bathysphere of Barton and Beebe may be lowered to far greater depths than it has gone, and that seems incredible. If remnants of prehistoric life live in moderate depths, where they can without difficulty come to the surface, they should by now be well known as to form and provided with scientific names, as are whales, sharks and the general run of sea creatures. Science has pretty thoroughly explored the ocean. Little seems to have escaped the sharp eyes of oceanographers. It has been unusual for a number of years for any new species of large size to be added to the scientific lists. And with all this search, there has been no conjecture, no rumors of puzzling "monsters" among scientific observers.

Ask these technicians, who have written volumes

about the oceans' life, to account for the tales of sea "monsters," and they will mention several known forms which continue to support the belief. They lay the chief blame upon the giant squid, a creature which attains a length of fifty feet, its long tenacles carried before it, two of them heavily knobbed at the end, and when simultaneously raised looking like the neck and head of a great serpent. Some of the giant sharks display antics which, from a distance, might puzzle a lay observer. Then there is the great leathery turtle, which grows to a weight of a thousand pounds, its head and shell unlike those of the familiar turtle. There is the strange elephant seal of the Southern Hemisphere. It goes on migrations, attains a weight of several thousand pounds, and can inflate the top of its snout or extend it like the foreshortened trunk of an elephant. The head of such an animal when swimming would be a fearsome sight to an observer not knowing what it was.

Thus the belief in sea "monsters"—and only the ocean is credited with harboring such mysteries—may be nothing more or less than a hangover from mediaeval superstitions on a par with vampirism. The French naturalist, F. A. Pouchet, of the Museum of Natural History at Rouen, wrote on the subject of alleged marine monsters in 1867, and convincingly described the foundation of such rumors. He stated that the people of antiquity had many superstitions, which became widely spread in the Middle Ages, "a period of simple ignorance and ardent faith." These were the days of general belief in magic. From continental Europe stretched a seemingly endless and mysterious sea. Land monsters, stated to lurk in the Alps, had been described. With the conquest of terra firma they failed to materialize. The vastness of the oceans remained, and horrible tales of sea serpents accumulated to satisfy public credulity. Scandinavian writers produced a composite description of a sea serpent six hundred feet long, with a head resembling that of a horse, rearing itself as high as a ship's mast and giving vent to a hissing like the sound of a tempest. When the fishermen encountered it, they sailed in the direction of the sun, so that the bright light would blind the monster.

Even in the works of the early naturalists sea creatures are credited with enormous size. Buffon states that in the northern seas there are cuttlefish of such a size that a whale is a dwarf compared with them, and that when motionless and half out of the water, their bodies coated with marine plants, they have been mistaken for floating islands. Some of the old chronicles relate that sailors have been deceived by such treacherous signs, and have been known to row up to and land upon the backs of such monsters.

The disappearance of ships was attributed to the

attacks of monsters, which allegedly existed in a wide variety of forms.

As the days of belief in mediaeval superstitions receded, changing to the milder fanaticisms of the Salem witchcraft type, the conjectured size, ferocity and abundance of sea monsters likewise decreased; but a remnant of the belief remains today. And so do ideas about walking under a ladder and the fateful significance of number 13.

Turning to the appearance of the "monsters" of 1933 and 1934, we may check off the apparition of the northerly Pacific coast as remaining a mystery. There are no further reports of it. The "monster" washed up at Querqueville on February 28, 1934, described as having "a head like a camel," was identified by scientists as a species known as the Basking Shark, its gill structure so decomposed that only the solid bones of the head remained, for the moment deceiving investigators. Its stranded carcass is shown on an accompanying photograph.

The Loch Ness "monster" remains a lively mystery, supposedly still lurking in the elongated lake. It first appeared in the summer of 1933 and continued to be seen into 1934. Loch Ness is a large body of water, about twenty-five miles long, averaging over a mile wide and in parts seven hundred feet deep. It is in the northeasterly part of Scotland, and is connected by a river with the sea.

MARINE IGUANA

This coal-black creature, living on sea-washed rocks of the Galápagos Islands, imparts a fair idea, in miniature, of what prehistoric monsters looked like.

A SEA SERPENT

Despite stories of colossal marine serpents, this type of creature, with paddle-like tail, actually a sea cobra, and attaining a maximum length of eight to ten feet, is the only proven "sea serpent."

*Permission of the Illustrated London News*

THE LOCH NESS MONSTER

Drawing made by G. H. Davis, special artist of the Illustrated London News.

That "something" played around Loch Ness during this time is amply proven by authoritative observers, at least fifteen of whom made drawings of it at distances varying from a hundred yards to a mile. However, it was seen much closer, once as close as ten yards by moonlight off Temple Pier at Drumnadrochit. I am indebted to the *Illustrated London News* for details of the creature's actions. It was seen during summer and winter months, and once during a snowstorm. It was described as equally at ease during smooth or rough water, and swimming as fast as a lake steamer. Some of those who sketched it noted it rearing a head and neck at least five feet above the water. In most of the drawings the body is only indicated, or not portrayed at all. Others show an elongate body, hinting of stout, serpentine outlines, but with *vertical* undulations protruding from the water, which would imply swimming motions never assumed by serpents. The *Illustrated London News* sent their special artist, Mr. G. H. Davis, to the Loch "that he might record there the evidence of a number of reliable persons who are convinced that they have seen a strange apparition in the waters of the Loch. . . . Our artist's method was to draw the monster from a sketch drawn by the witness concerned; and his drawings were passed as correct by the witnesses."

On an accompanying plate is Mr. Davis' drawing, reproduced by permission accorded me by the *Illus-*

*trated London News*, which carried a caption explaining: "The 'monster' depicted above is that vouched for by Mr. B. A. Russell, M.A., of the School House, Fort Augustus. Mr. Russell, a very calm, level-headed man, states that he saw the mysterious head, as here drawn by himself and by Mr. Davis, on Sunday, October 1, 1933, between 10:00 and 10:30 in the morning, and that it was visible for twelve minutes. His observation point was near Captain Meiklem's house, near Fort Augustus, at a height of over one hundred feet. The sun, he told our artist, was shining and the Loch was absolutely calm. He watched the head and neck moving with a 'horizontal undulation' about five feet above the surface, and noted that the creature covered half a mile during the twelve minutes he had it under view. It was some seven hundred yards from him and was silhouetted against the pale gray water, so that it was very evident."

The significant point about the Loch Ness monster is that reports of its size from reputable observers were not toned down but *grew.* A sketch resulting from the observation of the Misses Rattray and Miss Hamilton, near Dores, compared its size to the steamer "Grant Hay," in the wake of which it was following. The steamer is eighty-six feet long and the monster was estimated as being three-quarters the length of the vessel, judged from the humped ridges of its body showing from the water.

In answers to American newspaper queries about the Loch Ness apparition, I have stated my conviction that it cannot be a reptile. I know those lakes of Scotland. I have traveled past the impressive Grampian Hills in midsummer with a tightly buttoned coat, and welcomed the sight of a friendly inn and a crackling fire. The only reptiles that can survive there are the little grassnake and the adder, in sheltered hollows of the moors where the sun strikes in. When the cold mist presses down they have their snug retreats between strata of soil. I have traversed Loch Katrine and Loch Lomond—well buttoned up, in August—and those lakes are south of Loch Ness waters. No, Loch Ness would benumb a reptile. And remember, this "monster" was seen to disport itself during a snowstorm!

As to actual marine sea serpents there are many kinds, oceanic specializations related to the cobra, that inhabit the tropical Pacific and Indian oceans. About fifty kinds are known. Some swim in schools. Most of them are three to four feet long, with the tail vertically compressed like an oar. They are sometimes to be seen a thousand miles from shore. A few attain a length of ten to twelve feet, but they have slender necks and heads and are not to be rated as "monsters." All are well known to science. Not one among their number could survive more than a few days as far north as the English Channel. Place one of them in the cold waters of Loch Ness, and within an hour, it would be numb and helpless. They require a water temperature of seventy-five degrees or more. The activity of a reptile depends upon a moderately high temperature of the air or water surrounding it. So from this standpoint the Loch Ness monster is eliminated from reptilian relationship.

What can it be? Its existence is indisputably established. My deductions, after studies of the larger of the world's wild creatures including many marine types, point to the probability of its being a large seal of some kind. I base this belief on the following points:

There is slight possibility of the creature having long existed in Loch Ness, or it would have been previously seen; hence it worked its way in from the sea, which it could readily have done, through the river connecting the loch to the coast. Loch Ness is a body of fresh water, and comparatively few marine creatures are adapted to changing their habitat from salt to fresh water. Among seals, however, this has not infrequently been noted. The small harbor seal of the northeasterly seaboard of the United States has been seen—with some astonishment on the part of local residents—cavorting in the waters of Lake Champlain. On the Pacific coast seals have been seen in lakes a hundred miles from the sea.

Two species of seals, both of which occasionally occur along the Scottish coasts, might have provided a particularly large member that decided to go pioneering into the waters of the loch. One is the gray seal, which attains a length of ten feet and a weight up to five hundred pounds, and the other is the North Atlantic hooded seal, almost as large, but more unique in being able to inflate the top of its head into a grotesque shape. Occasional male seals far surpass the average measurements of their clan, and might be mistaken for unusual "monsters," if seen from a distance.

My other deductions pointing to the Loch Ness apparition being a seal, come from the speed at which it is seen to swim, and the fact that it continued to remain in the loch. Seals feed largely upon fish, and the loch is well supplied with such food. Also, a composite view of the drawings of the "monster" indicates the head of a seal.

A ten- or twelve-foot male of the gray seal—and the latter measurement is not improbable in view of occasional "record" specimens among other animals—would raise a heavy wake in its trail as it ploughed the water. The gray seal is a rapid swimmer and sometimes pursues an undulatory course through the water. This might result in "follower waves" behind the creature, which, under certain conditions of light, might delude an observer half a mile away on the shore into thinking that the wake was part of the animal itself, thus accounting for the monster's being alleged to appear to be three-quarters the length of a lake steamer.

After no word from the Loch Ness "monster" in several months, I received a telephone call from Universal News Service in June of this year, asking me if I knew of an animal called a Regalecus. They said that the body of a marine creature had been washed up on the shores of the Moray Firth, a large indentation of northeasterly Scotland, into which flows a river connecting with Loch Ness. The creature was said to be twenty feet long, and fishermen called it a sea serpent, suggesting that here was the solution of the Loch Ness monster. Scientists called to investigate it pronounced it to be a Regalecus.

The Universal News had dug into encyclopaedias and couldn't find any such animal. I told them I had never heard of it and referred them to the director of the New York Aquarium, as the creature was evidently of marine origin. The officer at the aquarium recalled *Regalecus* as the scientific name of a rare fish. Its full name was *Regalecus glesne* and its natural home appeared to be in subtropical waters, from which at times it might be lured to follow warm currents flowing northward. Wandering out of these fickle ocean paths it would become chilled and helpless and be cast up on some northerly shore. It was remarkable for its serpentine body, vertically compressed, with a high crest of fin along the back, higher near the head. In appearance it was a "plumed serpent," scientifically listed and in no way a mystery, its gills adapted to sea water and

*Permission of Wide World Photos*

THE QUERQUEVILLE MONSTER

Washed up on the coast of France and said to have a head like a camel. It was later identified as a type of shark with gill structure disintegrated and only the head bones remaining anteriorly.

A PET MONSTER

In the summer of 1934 the Griswold-Harkness Expedition returne
from the remote island of Komodo, in the Dutch East Indies, with thre
of these "Dragon Lizards." They are over eight feet long.

not fresh. It would have quickly perished in the cold waters of Loch Ness.

I predict that no vertebrate sea monster, unrelated to the known clans of sharks and other big, gilled fishes, will ever be discovered by man, to the definite extent of achieving a standing, scientific name. But tales of the sea serpent will continue to thrive so long as mankind sails the immensity of Old Ocean and remembers its legends.

The foregoing is, in effect, what I said during a public lecture. Shortly after the lecture, in fact in mid-June of the present year, two young explorers, Lawrence T. S. Griswold and William H. Harkness, returned from the remote little island of Komodo in the Dutch East Indies, east of Java, between Sumbawa and Flores. They brought three of the giant monitors which, strangely enough, appear to be confined to Komodo. These reptiles are so disproportionally huge as compared to other lizards that they have been called "dragon" lizards. After traveling over twelve thousand miles in teakwood crates they sprawled their respective eight feet in a sunny yard outside the reptile house, reared their massive heads and darted greenish, forked tongues a foot long. They looked like restorations of prehistoric creatures, with the whip-like tongues supporting the dragon title. Newspaper stories commented on this, and it was alleged, though where the statement came from I don't know, that the "dragons" could claim relationship with *Tyrannosaurus rex*, a giant dinosaur. Tyrannosaurus was about the world's worst prehistoric monster. It arrogantly stalked around on its hind legs, towered to a height of about twenty-five feet, and had teeth the length and curvature of half the blade of a hand scythe. The article also explained that the Komodo "monsters," had suddenly been discovered, to the astonishment of the scientific world, in 1916.

Here is a case of how a scientist can get into a jam. A man who had been at my lecture telephoned, and with barely veiled sarcasm asked how I explained my denial of the existence of "monsters," when a pair of creatures called dragons had actually arrived at the Park, were only recently discovered by scientists, and related to the truly awful Tyrannosaurus.

I patiently explained that the "dragons" were monsters only when compared to the size of the general run of lizards, that a medium-sized crocodile—also a reptile—would be larger and heavier, and that they belonged to a well-known genus of lizards, of which one common species found in India grew to two-thirds their length, although much lighter in build.

When the party on the telephone reverted to the reports about the "dragons" being related to Tyrannosaurus, thus hinting of other living remnants of prehistoric life, I told him that I had said nothing about such a relationship, and explained that the last member of the Tyrannosaurus clan had expired a generous number of million years ago, and that the bones of those giants existed only as rock-bound fossils. But the man signed off with a satirical "Good-bye."

And that was not the end of it. I answered similar questions. Possibly I convinced a few, but many who read those stories and noted the alleged relationship with what palaeontologists call the Tyrant Dinosaur, are ready to absorb a new crop of "monster" tales.

That is the way it goes. A scientist must watch his step all the time. The checking of the vampire paper illustrates how he must guard against a comeback from intently watching scientific colleagues, and my denial of the possibility of "monsters" indicates how such an explanation must be handled with a thought of not puncturing too roughly pet theories of the public.

The rate of a scientist's progress at times becomes so mincing and tip-toe, that occasional breaks for complete change and rest are made by all members of the fraternity. I can remember one of these, the past year, leaning over the rail of a ship, bound for the tropics. For a half hour I had been idly watching the bow-wash roll out beneath me. Turning, I saw another, elbows flat on the rail as mine were. The figure was of a scientific professor of the Princeton faculty. We exchanged greetings.

"Where bound?" I asked.

"Just taking the circuit with the ship," he answered. "Gives me a rest of three weeks—sea air and a change."

I told him I was dropping down to Panama to spend my vacation.

"Thought this would be a good chance to read over some manuscript. Getting out a formal paper this fall," he said.

"I'm going up the Chagres valley to check the occurrence of *Lachesis*," I conceded.

Two scientists relapsed into silence, elbow to elbow on the ship's rail. One was ruminating over a manuscript, the other figuring on field paraphernalia. We had sought rest from the exactitudes of our professions.

CHAPTER XV

# THE NEWS POINT OF A ZOO STORY

I HAVE often been asked about the publicity department at the Zoological Park. The question is natural and is prompted by the many news stories that have appeared. My answer has been that there is no such department. I explain that we have a publicity committee, of which I am a member, but so far as I can recollect, that committee has never had a meeting nor discussed anything. This explanation produces surprise, for the Park does get a large amount of publicity. A more complete answer makes an interesting story.

Several factors contributed to publicity about the Park taking an early hold. In the first place, the institution was opened, approximately thirty-five years ago, under the formidable title of the New York Zoological Park. That is its name today, but owing to writers' stumbling over the long title, the easier term of Bronx Zoo was coined and used in earlier articles. The shorter title is now internationally known, although the older members of the executive board still shudder at it. But the formal title attracted attention in the early days to something *new* in the way of a zoological collection, and anything new, in a big way, is of interest to the press.

When the Park was first opened with only a few buildings, we could do little more than tell about the elaborate projects in sight. But the blue-prints of large buildings soon to be erected suggested stories. They indicated a combing of the far corners of the world for rare specimens, animals never before exhibited alive. This stirred up an interest in watching for the arrival of such specimens. As my job before coming to the Park had been on the staff of the *New York Times*, friends I had made in newspaper work came to see me. I had acquired a nose for news myself, and was able to talk in an understanding way about animals and the plans for development, so a small clan of newspaper men came to depend upon the Park for "special stuff," and got it. No mere "menagerie" could have provided such information.

The New York Zoological Park, as its heavy title indicated, was treated like a great museum of living exhibits, each with a fascinating background of habits. Adventures had accompanied the capture of the specimens. This was new, hence news to the extent of being worth special Sunday stories and magazine articles that went into considerable detail. Accompanying this increasing interest in the completion of new buildings and the arrival of specimens, was the development of the

Park's photographic department in the charge of Elwin R. Sanborn. Its purpose was the establishment of an extensive file of scientific records. Mr. Sanborn's work was soon recognized as having a "style" in animal portraiture quite unique in its excellence. Permission was sought and accorded to print these pictures in descriptive articles. Here again, in this animal portrait gallery, was something that was new. It was not many years before these pictures were being produced in various parts of the world. Hence interest in the Park continued to grow until it became recognized as a rich source of interesting stories.

We are visited regularly and frequently by our newspaper friends. There is no need to invite them or solicit publicity, hence no need for the publicity committee to function,—which it doesn't. It is rarely that a story is *sent* to the press. Infrequently, once or twice a year, I may call up the New York papers on the telephone and explain the details about something which in my opinion would be particularly interesting to the public. Such informal calls to the city desk have in each instance received a uniformly cordial response over a long period of years.

What are the favorite topics for stories sought by visiting reporters?

In the old days, when one rarity after another was coming in, the most frequent question was about what new kind of animal had arrived, or was on the way.

And those were the days of big stories along that line. I can remember when Carl Hagenbeck sent an expedition into remote Mongolia for the alleged wild, ancestral form of the domestic horse. The species, never before exhibited alive, had been named in honor of a Russian scientist, and hence bore the title of *Equus prjervalskii.* Several specimens were captured, three landing in America and coming to the Park. There was a story at that time which told about the difficulties of the expedition; and a year later, when a colt was born, the infant was thoroughly photographed by press photographers. Hagenbeck provided us with several such celebrities, among them the pygmy hippopotamus discovered in the Golah district of Africa, and the pygmy elephant. The pygmy hippos made a particular hit as they were no bigger than a hog, and when compared with Pete, the Park's Nile hippopotamus, showed about the same relative proportions as a locomotive and a handtruck. One of the points brought out about the pygmy hippos was that some scientists had scoffed about there being any such animals. There were frequent requests to pose Pete on one side of the fence and one of the new arrivals on the other, because he made such an excellent standard for comparing sizes.

The pygmy elephant was immediately adopted by a big Indian elephant, and noting this the press photographers hit upon the idea of having the pygmy stand beneath her. The massive but good-natured Alice was posed many times in this position.

Another of the news hits from Hagenbeck was a hybrid or cross between a lion and a tiger. This was a huge and savage-looking animal, and representatives of the press unsuccessfully sought to induce us to say that this combination animal could "lick" any lion, or any tiger in existence. As a news point, that feature of the animal fell flat. Incidentally it was exceptionally good-natured with the keepers, and blinked in unconcern at many photographic flashlights.

An adult mouse deer, standing barely a foot high, attracted much attention as the smallest known species of deer, but it invariably hid in the hay when photographers approached. We spent a lot of our time talking with photographers about the need for patience when getting a timid animal's picture. But the smallest kind of deer in the world, and the only one in America, was a novelty, and nobody would go home for fear someone else would get a "beat" on the picture. We finally compromised by saying that we wouldn't allow anyone to remain so long as the animal sought to hide, and that photographic possibilities would be left to Mr. Sanborn's department, who would simultaneously release prints to the press, if photography was found to be practicable. But the animal remained so timid that we arranged a deep and sheltered place for it, and it was never photographed.

Shortly preceding the outbreak of the World War, Mr. Robert C. Garner, commissioned by the New York Zoological Society to go to Africa and obtain a gorilla, cabled that he was on his way to Europe and had a young gorilla. At that time no gorilla had been successfully exhibited in the United States. Some visiting reporters were told about the successful quest and quite a story broke, built up largely on Mr. Garner's life study of the vocabulary of monkeys. The arrival was awaited with great interest, as it was intimated that the gorilla was being cared for like a human child. Then the war broke and the newspapers were packed with exciting news. Americans seeking the home country were jamming every available steamer. The inference would clearly be that if Mr. Garner managed to reach America with his gorilla at all, he would arrive with barely a paragraph. But he sent a cable from the other side, which swung him and his gorilla right into the middle of the war news. He cabled that he had found accommodations on a steamer, a small cabin for himself and a tool house on deck for the gorilla, but added: "I am occupying the tool house and giving the cabin to the gorilla." His humor and the comedy of the situation made such a hit that his arrival was written up in fair detail.

In Australia is a cunning little animal the size of a cub bear, looking in fact like a furry "Teddy Bear." It is not a bear, but related to the opossums. This strange-looking beast, called the Koala, feeds upon the leaves of the eucalyptus tree. Ellis Joseph, the well-known animal dealer, escorted one of them all the way from its home country. It was the first ever to arrive alive in America. The press was notified by telephone.

The arrival of the Koala failed to make much of a stir. True enough it was rare and its pictures looked cute, but in answer to questions about its habits we could say little. The leaves brought by Mr. Joseph had given out a day or two before the ship got in. He experimented and found that his bear would eat moistened graham crackers. On its arrival we begged a branch from an exhibition eucalyptus tree in the great glasshouse of the Botanical Garden, and promptly wired the Pacific coast, where eucalyptus trees have been imported and thrive, to put a load of branches in a refrigerator car and start them for New York. Before they arrived we had nearly stripped the tree in the Botanical Garden.

A reporter casually asking how the Koala was getting along was as casually given these details. He wrote a "funny" story about the exhibition tree and refrigerator car being involved in this animal's eccentric appetite. It was a "beat," though not one to cause any hard feelings. However, it again directed attention to the Koala, which died before the car arrived. Our unsuccessful attempt to save this animal proved to be the real point of interest about it to the newspaper men.

Three arrivals, above all the others, stand out as producing the most widely circulated stories in the history of the Park. The first was the Australian Duck-bill or Platypus, another of the Australian rarities brought over by Ellis Joseph. The high points around which the stories about the Platypus were written were the price we paid for the animal—two thousand five hundred dollars for a specimen no bigger than a musk-rat—the fact that it was the first of its kind to leave Australia alive, and that it had the body of a four-legged animal but the beak of a duck. As if these high points were not enough this four-legged monstrosity laid eggs, hatched them in a nest, and then reared the young with milk. It was so temperamental that it could only be exhibited one hour each day, and as its reputation grew long lines of visitors kept forming to look at it. The stories spread all over the world and even flashed back to its native country, Australia.

On a par with the duckbill stories were the descriptions of the "dragons" brought from the remote island of Komodo by the Douglas Burden Expedition. These great lizards, over eight feet long, and said to have relatives on the island twelve feet long, had only recently been discovered, and were alleged to have inspired the outlines of the dragon on the Chinese flag. Visitors seeing them after their great acclaim might have been disappointed. To us, however, and to visiting scientists, they were astonishing in comparison with the size of ordinary lizards, and our enthusiastic summaries about them were eagerly noted and amplified by the press.

Third, but not least as a star animal, was the vampire bat, a recent exhibit elsewhere described in detail in this book. The news point in this instance came not only from the fact that this was the first vampire ever to be seen publicly, but also because the very term *vampire* carried a thrill associated with eerie places and sinister doings.

Apart from stories built about the arrival of rarities, there has been a wealth of other events of interest to the press. Stories have been written about accidents, serious and humorous incidents, scientific discoveries, prospective expeditions and the results of expeditions. There has been a fair number of accidents, as is inevitably the case with a big collection of animals, but, fortunately, few have ever related to visitors. The escape of an animal involves publicity we very much dread. We have had a few escapes, but no visitors have been harmed. Yet it was the escape of an animal that produced *the* biggest story that ever appeared about the Park. It ran to columns' length day after day for over a week. What made the story big was that it concerned a puma or *mountain lion.* The latter title was mostly used, as it sounded more snappy, although it was varied with cougar, panther and catamount. We shuddered when we considered the effect upon the public.

The affair flashed out of a clear sky. A young puma arrived. It was two-thirds grown, and its traveling cage had been made with slats of oak instead of bars, because the animal was allegedly tame. The cage was placed in a hallway near my office, for it was time for lunch and I wanted to get the meal out of the way before studying where the puma should be placed permanently. I remember filling a water pan and shoving it into the cage, thinking that the express company might not have given the animal a drink for several hours. When I left, the puma was noisily lapping the water. I turned from that cage with not the remotest thought of worry. The animal had traveled in it a long distance, occupied it in a railroad car and delivery trucks.

When I returned from lunch, with an idea of where the puma would go—a cool enclosure under a spreading oak—I made a disturbing discovery. Two of the slats of the cage were broken. The puma was gone!

I sought the keepers of the animal headquarters building. Had they seen a puma strolling about? They had not. I did not entirely lose hope. Had someone hacked out two of the slats and taken the animal somewhere? The answer was negative. A puma was at liberty in the Bronx! It is not our principle to hide things. If anything happens we face the music. We conceded that a puma had escaped. The story was passed around. All the morning papers carried it.

At the time when this happened United States sheriffs and deputies were on the trail of an outlaw who had acquired fame for his evasiveness. His name was Tracy. The only thing the newspapers did that was not in accord with the information we gave out, was to name the puma "Tracy." The whole Bronx joined in the search. Nervous mothers kept their children indoors when the titles of mountain lion and panther were applied to the animal. As the fruitless search went on, the stories grew. There were reports of picnic parties being raided, sandwiches being gulped down by a hungry beast, picnickers in full flight. We insisted that the animal was tame. Outside observers described a raging beast. There were no casualties. About the fifth day a cartoonist produced a double column drawing indicating the two points of view. A beast the size of a lion, with bared fangs, was shown as one version. As ours, there was a kitten-like creature with a ribbon bow around its neck. After columns of publicity, the puma was brought back, following a telephone call, by Herman Merkel, Chief Constructor of our staff, who drove out to a farm in the Eastchester district. He found the animal temporarily adopted as a pet, and brought it back sedately sitting on the seat beside him. This demonstration of docility killed the story then and there.

Another big story was the attempted theft of a cageful of cobras. We found a strange contrivance for boxing them, and a hole cut in the steel door of the cage presumably through which to fish them out. Here was the Park's star mystery story, and I suspect that detectives working on the case were hard pressed for theories. We figured that the would-be thieves had been frightened away by the night watchmen. My supposition that no weird murder was contemplated fell flat. I suggested that it was more probable that the snakes were sought by a vendor of "snake oil" for ballyhoo purposes. I fear I disappointed our newspaper friends in not substantiating a better point for a story.

Eccentric animal traits have provided many stories, and I confess that in many instances I have pondered how best to make such traits stand out so as to provide good news points. It is the old instinct of the newspaper man, who feels guilty in seeing good stuff go to waste. The orang discovering how a stick could be used as a lever, then breaking away his oaken trapeze bar and using it as an effective instrument for bending the bars on the front of his cage, made a good story of animal psychology. A bear that escaped, raided a pie counter, then climbed back into his open-top den, made another. The discovery by some mountain goats that they could run up the leaning trunk of an immense oak tree, thence leap nimbly from one bough to another, until they were silhouetted against the sky, produced interesting reading, novel photographs, and a follow-up by the topical movie people, who produce animated news reels for the screen. Such incidents are clearly indicated as stories. But frequently a reporter drops into my office, and asks if I have anything in mind for a story. I may not have anything immediately in mind, and start to think. Some incident is recalled. At first thought it may seem weak in a news point, but warranting investigation.

An example of such was the frequent bellowing of the colony of big alligators in the reptile house. Day after day the water in the great tank quivered during such demonstrations. At times it was difficult to use the telephone. The bellow of an alligator is a mating call. Its reverberations roll along the ground. In past years the alligators had bellowed infrequently. What was the reason for all this noise? The item was interesting, but had no particular news point.

A reporter and I talked to the keepers. The men suggested that the alligators were stimulated to bellow by blasting operations several miles away in the northerly part of the city, where a tunnel for a new water system was being blasted deep in the ground. I told the reporter to hold the story over for a day, that I would check the suggestion the next morning and he could learn the result by telephone. The next morning I went into the deep cellar of the reptile house, where everything was quiet, and waited to see if I could detect the distant blasting. It came at the time the keepers insisted it occurred daily, in the form of faint thuds, which I figured would produce waves of ground vibration apparent to sensitive forms like the alligators. Even as I was thinking about it, the rumbling bellows from the 'gator tank above were starting. Hence a fair story followed, amplified by my trouble with the telephone.

Another instance of something coming from comparatively nothing was a request during a dull period for something doing in the line of news. A small Australian animal had died that day. It was rare, but of a type that had little exhibition value. The report of its death would not have warranted a paragraph. I happened to think that members of the group to which it belonged, while no bigger than a terrier dog, had a vermiform appendix up to six feet in length, but apparently never had any trouble with it. Short items of this kind may attract wide attention.

Between arrivals of unusual specimens from remote places, which form the points of the longer stories, the minor incidents crop up: The post-mortem of a sea lion reveals ten quarts of round stones in its stomach, a nickel whistle, a rubber ball and a handkerchief with an embroidered monogram; an elk is the proud mother of a fawn and the news point about the episode relates to her having previously reared fifteen fawns in nineteen years; a fer-de-lance gives birth to sixty-eight young, showing the reason for this formidable snake being so common in the tropics.

I remember one of our past administrative officers giving a luncheon to members of the press who had written about the Park, the idea being to thank the reporters in a body for the care they had taken in accurately presenting news from the institution. Thinking that he would give them some dignified items upon which to base a story, he spoke about fossil bones recently unearthed near New York, recent laws relating to the importation of plumes from tropical birds, and the "bringing back" by breeding in protected areas of important types of animals that had been threatened with extinction. It was an instructive talk, but the boys didn't figure it as up-to-date news. They listened respectfully, but after the luncheon wandered around and happened to visit the bear keeper, Peter Romanoff. Here they learned, quite incidentally, as Romanoff started to excuse himself for a moment and climb a steep ladder to the top of the bear den ledge, that the keeper, whose parents had been Russian peasants, had a row of bee hives at the top of the ledge, that the bees had to go miles to get most of their honey, and in hurtling back through the air, buzzed within inches of the bears which sit for hours at the highest point of the rocks. Ivan, the thousand pound Alaskan bear, had selected as his favorite napping place a rock within a few feet of one of the hives, where the congregating, incoming and outgoing bees, caused a loud and steady hum. That was the story which followed their visit, stressing the point that Romanoff had never seen one of his bears stung by a bee.

The Park has acquired a reputation for strict authenticity in publicity. Possibly that is the reason why so many newspaper writers visit us. They feel that no tricks are carried up our sleeves, that we have not solicited the interviews. An occasional item may become distorted in some of the papers through the injection of snap into the information to make lively reading. Ninety per cent of the stories, however, are correctly stated and contain instructional points worth remembering.

We respect this interest of the press, seek never to be flippant, nor put anything over, as the saying goes. Yet I have seen a few items appear about which I have felt guilty, because I have unconsciously stressed some point, which by inference suggested a story. Coming hurriedly through the monkey house one day I saw a well-known reporter talking to one of the keepers. Some excellent special stories have come about through such conversations with sympathetic animal attendants. I took it for granted that material for an article was being gathered and was about to continue on my way when the reporter stopped me with a grin.

"Look up there," he said.

On a shelf, side by side, were two orang-utans, holding hands.

"Our primitive ancestors!" laughed the reporter.

"Yes, the good old human habit of holding hands may have originated with that clan," I jokingly replied, and went my way.

The next day a paragraph story, but with a heading that instantly attracted attention, appeared on the paper's front page. The story made such a hit that it spread across the country from coast to coast in a decorative bracket as a front page feature. In a way it was not incorrect, as undoubtedly we inherit many habits from our wild ancestors, whatever they were, but that is not the kind of story which, played up, is helpful to the Park, as it is open to severe criticism by scientists on the ground that we are facetiously theorizing on animal behavior.

Occasionally some of the papers that present the news in lighter vein, or the topical news reel men, ask permission to try out ideas with the animals. If these are not of a disturbing nature, or flagrantly undignified, we are inclined to grant permission, having it understood that such stories make no claims about extraordinary behavior. I remember a Christmas tree being set up for the young orangs and chimpanzees, and it being wrecked during the photographing of the event. Also the experiment of seeing what the polar bears would do when presented with fish frozen in blocks of ice. Incidentally the bears batted the ice cakes all over the den with such energy that photography was difficult. I doubt if the bears had any idea of definitely breaking the cakes, which as a matter of fact failed to stand up under the treatment; but what with the breakage and the melting of the ice the bears did obtain the fish. Quite recently one of the more profusely illustrated papers requested permission to bring up a big birthday cake to Pete, the hippopotamus. This plan grew out of a story that had appeared about our oldest animals. The idea was to celebrate Pete's thirtieth birthday. It was explained that some scouts were to participate, and we reluctantly consented. The huge cake, with thirty candles burning, was set in front of Pete's quarters, and as the mild-mannered animal stood with his nose almost touching it, lured to the spot incidentally by a hidden bunch of carrots, a batch of flashlight bulbs recorded the scene. The scouts ate the cake and we figured that our dignity had not suffered, as little was done beyond demonstrating Pete to be thirty years old.

A magazine writer was keen to note what effect a full grown gorilla would have if it stalked through the animal buildings of the Park. No full grown gorilla has ever been exhibited. This African giant of the apes attains an adult weight of four hundred pounds. A two hundred pound man was dressed in a fair imitation of gorilla attire, and from his antics I gathered he had read a considerable number of the more lurid adventure tales. He was a fearsome object, and champed jaws with broad and wicked incisors and protruding canines. He incased himself in the hairy garment in the keepers' room of the monkey house, slipped the mask over his head, pounded his chest a few times for practice, and stalked forth along the line of cages.

The monkeys were but mildly interested. Their attitude was one of envy that a member from the ape section could thus parade through the building. In the elephant house there was marked curiosity and trunks were extended. Here possibly, they seemed to think, was a particularly opulent visitor who might shower fistfuls of peanuts. The author conducting the experiment walked along with a batch of note paper, which remained blank for lack of items of interest.

Now followed by a bevy of interested visitors, the replica of a giant ape stalked toward the lion house, pushed back the swinging doors and started along the series of cages of that two hundred foot building.

Before he had passed three cages there was pandemonium. Lions and tigers turned near somersaults and leopards leaped to the top ledges of their cages. Of all places to find surprise at such a figure!

We grabbed the "gorilla" on either side and hustled him out to the tune of snorts and growls. The reactions to this experiment were so arbitrary and unsatisfactory, that, so far as I knew, the story was never written.

In July of 1933, Lawrence Griswold, who with William Harkness made an expedition to the island of

Komodo and brought back the second batch of "dragon" lizards to arrive at the Park, dropped into my office for a word of advice.

After I had congratulated him on the wide acclaim resulting from the expedition, he said:

"We are planning another trip. What do you consider the greatest animal prize that could be brought back?"

"The giant panda," was my answer. "An animal that looks like a big bear that has dipped its head and forequarters in a flour barrel. It's a spectral thing, as it has black circles around its eyes, making them look enormous. It lives in high western China, and no live specimen has ever been brought out."

"Good," said Griswold. "We'll get a panda!"

"You will have to reconnoiter those mountain passes with an aeroplane."

"That will work out all right," said Griswold. "I've over a thousand hours' flying, all told."

"And if you get in there, there are two other animals worth looking for: the giant argali sheep and the Marco Polo sheep. The former grows to a weight of five hundred pounds; its horns five feet, along the curve."

"Quite an animal to manage," remarked Griswold.

"Your only hope with the panda or the giant sheep will be to obtain the young."

That these young men are going, is evident. Advance

stories have appeared bringing out the favorite news points in zoo writing. And when I think of the atmosphere behind that trip, the strangeness of those mountains, bamboo forests extending to a high altitude, the planning of crates and capturing devices, the spectral appearance of the panda and its alleged strange habits, sheep of which the horns alone weigh over a hundred pounds—when I think of planning stowage on an aeroplane of infants of the strange trio to be sought, there is a mental flashback to newspaper days. I feel like writing that story myself.